SPACETIME AND ELECTROMAGNETISM

D1280610

SPACETIME AND ELECTROMAGNETISM

An Essay on the Philosophy of the Special Theory of Relativity

by

J. R. LUCAS

Fellow of Merton College, Oxford, and Tutor in Philosophy

and

P. E. HODGSON

Fellow of Corpus Christi College, Oxford, and Tutor in Physics

CLARENDON PRESS · OXFORD
1990

Oxford University Press, Walton Street, Oxford OX2 6DP
Oxford New York Toronto
Delhi Bombay Calcutta Madras Karachi
Petaling Jaya Singapore Hong Kong Tokyo
Nairobi Dar es Salaam Cape Town
Melbourne Auckland
and associated companies in
Berlin Ibadan

Oxford is a trade mark of Oxford University Press

Published in the United States
by Oxford University Press, New York

British Library Cataloguing in Publication Data
Lucas, J. R.
Spacetime and electromagnetism.
1. Physics. Special theory of relativity
I. Title II. Hodgson, P. E. (Peter Edward), 1928–
530.1'1
ISBN 0-19-852039-5
ISBN 0-19-852038-7 pbk

Library of Congress Cataloging in Publication Data
Lucas, J. R. (John Randolph), 1929–
Spacetime and electromagnetism: an essay on the philosophy of the
special theory of relativity / by J. R. Lucas and P. E. Hodgson.
p. cm.
Includes bibliographical references.
1. Special relativity (Physics) 2. Electromagnetism. 3. Space
and time. 4. Causality (Physics) I. Hodgson, P. E. (Peter Edward)
II. Title.
QC173.65.L83 1989
530.1'1–dc20 89-28064
ISBN 0-19-852039-5
ISBN 0-19-852038-7 (pbk.)

Typeset using TEX by J. R. Lucas
helped by Mrs. Julie Long, Mark Munday and Prasenjit Saha

Printed in Great Britain by
Bookcraft (Bath) Ltd
Midsomer Norton, Avon

Introduction

We have two aims in writing this book. We want to introduce readers to thinking philosophically about physics and we want to argue for a rationalist rather than an empiricist philosophy of physics. The book originated in lectures we gave for the Final Honour School of Physics and Philosophy in Oxford. The choice of two lecturers from different disciplines was deliberate. We saw questions from a different point of view, tackled them in different ways, gave different answers, and disagreed with the answers the other gave. The last point was particularly important for the purpose of communicating to undergraduates the spirit of the philosophy of physics. Inevitably, years of school induce a belief that there are always right answers to every question, and that it is the duty of the student to listen to his teachers telling him what those right answers are, and diligently to remember and believe what they say. It is always a salutary experience to discover that Sir may be wrong, and to have one don disagreeing with another not only demonstrated that they could not both be 100% right but encouraged undergraduates to venture their own opinions and to realise that in some cases truths could not be learnt second-hand, but must be discovered by thinking them out themselves, and through successive failures to form an adequate view of the matter. Philosophy is not just the acquisition of true views, but an activity, and it is possible, by thinking philosophically, to reach philosophical conclusions which are less inadequate than those one held before: but it is only through the activity of trying out different approaches, and living with problems over a long time, that we can feel out the difficulties of a philosophical problem, assess the merits of different solutions, and decide which one we should adopt.

In the course of years the views of the two authors have crystallized and converged. Agreement has not always been reached, but too much common ground has emerged for there to be confrontations which undergraduates feel impelled to join in. It seems more sensible now to widen the debate by exposing our views to public criticism. In writing them up we have tried to cater for philosophers who know little physics and physicists who are not used to thinking philosophically. In addressing ourselves both to physicists and to philosophers, we have a double task. Physicists learn the concepts of physics by manipulating them, using them to solve theoretical problems and to carry out practical experiments. They become second nature in much the way that classical Newtonian concepts are absorbed and seem natural to the layman. But there are disadvantages in absorbing ideas entirely by osmosis. It is all too easy to absorb false ideas along with the true, and it is much more difficult to unlearn at a later stage doctrines and concepts which were never explicitly formulated or argued for, but just taken for granted. And, as we shall see, many derivations in the Special Theory rest on assumptions which are seldom made explicit or argued for. Although the physicist is unlikely to learn much new physics from this book, he will come, particularly in the passages where he disagrees with our argument, to know much better what it is that he knows, and why he believes it.

Philosophers in the present century are strong on thinking, but weak on knowing. There is a danger of an easy scepticism becoming a cloak for intellectual laziness. Many philosophers have had opinions on time, space and causality, but have not taken account of the experiments and hard thought that physicists have given to these topics over the last three hundred years. Physics has not yielded answers to all the questions that natural philosophy has traditionally been concerned with, and the answers it has given are still open to revision and evaluation. But a philosopher is not seriously committed to his subject if he is not concerned, among other things, with the nature of nature, and will not understand it well unless he has some acquaintance with the knowledge won by the experiments and hard thought of the physicists.

Neither of our aims can be adequately achieved in the compass of one book. A proper understanding of the Special Theory needs much experience of problem-solving and manipulating formulae which we do not attempt to provide: there are no exercises. Physicists, trained to operate within the confines of scientific inference, find it difficult to attune themselves to the more etherial canons of philosophical argument, and may well be impatient as we labour the obvious, adducing reasons why space should be regarded as uniform or linear functions as simple. To the disappointed reader we can only apologize. We hope they will find some other books more specifically and successfully addressed to their needs.

We have been torn in other ways too. We have been torn between the need to simplify and the need to elaborate. We are confident we have fallen between these two stools, and that specialists will justly criticize us for leaving too many loose ends untied and not noting all the caveats and special conditions they would have registered, while others will complain that they cannot see the wood for the trees. Again we apologize, but plead that it is only by addressing ourselves to disparate readerships, and having in consequence to adopt a different focus at different stages of the same exposition, that we can hope to bridge the gap between the two disciplines and the two cultures they typify. It is a gap worth bridging. It has greatly impoverished our intellectual life in the last hundred years. And it cannot be bridged unless we address both physicists and philosophers, scientists and non-scientists, together. When the focus is blurred, the necessary bifocalism of the enterprise is our main explanation and excuse.

We presuppose some prior knowledge on the part of the reader. There are many references to *Space, Time and Causality*, which one of the authors published four years ago, and which aims to provide an introduction to the philosophy of space and time. There are some differential equations, some matrices, some reference to hyperbolic as well as the ordinary trigonometric functions. Any one who has A-level maths or physics should be able to follow the symbolism, but of course, real understanding is harder to come by, and often it takes time and much thought to register the significance of an equation. Philosophers may well find the formalism unfamiliar, but it is essential for a serious discussion of the subject. Mathematics enables ideas to be expressed with a precision and clarity that cannot be attained by verbal facility alone. But mathematics *is* difficult, and we have therefore, though not for this reason alone, drawn back from the affine connexion, the metric tensor, and all the refinements of the General Theory, and sought generally not to be more reader-unfriendly than was required by the nature of the case.

Our choice of title is significant. Negatively, it avoids the word 'Relativity', because this word has given rise to many needless confusions. It has led many to suppose that Einstein had shown that all is relative, and suggests that the Special Theory and the General Theory are essentially similar. We deny both. Positively, the title takes up the theme of *Space, Time and Causality*, but with a difference, indicating the integration of space with time into Minkowski spacetime, and glossing causality in a particular way.

Einstein's two theories of relativity have been made needlessly difficult by their name. Although as a matter of history Einstein's two theories were called the Special Theory of Relativity and the General Theory of Relativity respectively, we shall eschew the word 'relativity' altogether, and refer to them simply as the Special Theory and the General Theory respectively, and consider them in comparison with Newtonian mechanics. There are thus three theories under discussion. Newtonian mechanics is a theory of material bodies, and the forces and impacts between them. The Special Theory is a theory of electromagnetism, and the propagation of causes. The General Theory is a theory of gravitation and free fall. Although Einstein's break with Newtonian metaphysics was of great importance, our general argument will be that the Special Theory is much closer to Newtonian mechanics than to the General Theory.

That *Space* and *Time* should be integrated into a single *Spacetime* is the great insight of the Special Theory, and leads us to regard Spacetime as one of our fundamental categories in terms of which we make sense of the world around us. But it is not the only one. Causality is equally important and, at least as far as the Special Theory goes, not to be subsumed under a fundamentally geometrical form of explanation. *Spacetime and Causality*, although generally apt, would fail to catch the peculiar feature of the Special Theory, that the propagation of causal influence is by electromagnetic radiation. The light-cone is fundamental to the topology of Spacetime. To the layman our theme is Spacetime and Light: the physicist knows, however, that light is but a special form of electromagnetism, and our title acknowledges the importance of this discovery in our search for a unified theory of all the phenomena of physics.

The Special Theory is far from providing a key to the whole of physics. Though we touch on its relation to the General Theory and Quantum Mechanics, we have nothing to say about Nuclear Physics or Quantum Electrodynamics. It may be that they are related in ways we are unable to discern. Perhaps a future Lord Cherwell will smile at our obtuseness.[1] Certainly we have been impressed by the

---

[1] See below, §10.7.

way different concepts and different arguments hang together and combine to yield by different routes the same conclusions. The design of the book-jacket, taken from Figure 5.1.1, is intended to convey our impression of the interrelatedness of things.

We have organized our thoughts around a theme that runs counter to the empiricist fashion of the age. We are at pains to stress the rationalist strain in our understanding of the physical world. We are rebels against the Logical Positivism that still pervades much thinking about the philosophy of science. We believe in both the rationality of natural science and the reality of the world it studies. We are experimental rationalists, believing that there are many modes of reasoning we do, and properly can, employ in thinking about the natural world, but that we may always fail to reason aright, and must therefore be always ready to submit our conclusions to experimental test. We are led therefore to reverse the order of argument. Instead of clear-cut derivations familiar in text-books, we are often working backwards, trying to articulate the know-how of the working physicist, and to make explicit the principles that guide him in the exercise of his "physical intuition". But there is no assured end to this process of formulation, and we know that, again, to some of our readers we shall be labouring the obvious, while others will feel that we are taking too much for granted, and may well be smuggling in unwarranted assumptions. But it is inherent in the enterprise we have undertaken that we shall lay ourselves open to such criticisms; and if we can sting our critics into doing better what we have done imperfectly, we shall be well rewarded.

It remains for the authors to acknowledge their debts to many colleagues, and in particular to Harvey Brown, and also to each other. In addition to the fruits of collaboration that can be expected from a joint enterprise, it is worth recording that the philosopher has learnt much philosophy from the physicist, and the physicist much physics from the philosopher.

Oxford   J.R.L.

1989   P.E.H.

Plan

In the first four chapters we adopt four different approaches to the Special Theory. In the first chapter the theme is ontological: the Special Theory is a theory of a unified spacetime, which is seen as the fundamental reality in place of the disparate space and time of Newtonian mechanics. In the second chapter the theme is sameness: the fundamental question is "In what frames of reference are physical laws the same?", and the equivalence between them is defined in terms of an insight due to Galileo and refined by the theory of groups. In the third chapter the underlying relation is not that of sameness but that of order, and the Special Theory is developed on the basis of a certain ordering relation that can, with qualifications, be identified with that of a possible cause. Communicating is in many ways like causing, and in Chapter Four a "Communication Argument" is presented and assessed.

Each of these approaches can yield the Lorentz transformation (although we do not spell this out in detail in Chapter Three), which is the characteristic transformation of the Special Theory, and in Chapter Five we consider the significance of there being many different derivations all leading to the same conclusion. Having established the Lorentz transformation as the keystone of our argument, we use it to give an integrated view of mechanics and electromagnetism and to derive Maxwell's equations in Chapter Six. Chapter Six carries through into physics the integrated spacetime geometry of Chapter One, and shows how the historical argument from Maxwell's equations to the Lorentz transformation can be made to go in the opposite direction.

Chapter Seven investigates a different sort of sameness from those considered in Chapter Two. In Chapter Two the samenesses are expressed by continuous transformations: displacement, re-orientation, and uniform velocity. In Chapter Seven the transformations are discrete ones, reflections and their generalisation in changes of parity. These offer an *aperçu* into a strange, looking-glass world, in which baffling ideas of charge conjugation and reversals of the direction of time turn up.

In Chapter Eight we review the Special Theory both historically and in comparison with Newtonian mechanics and the General Theory. In some ways the Special Theory marked a break with its predecessors, but in many respects it emerged from the same ideas and principles as had Newtonian mechanics and its predecessors. The similarity between the Special Theory and Newtonian mechanics becomes clearer when they are both contrasted with the General Theory. The implications of the General Theory are not yet fully understood, and no considered assessment is offered. We view it only from the over-simplified standpoint of the Special Theory, so as to provide a foil to our understanding of the Special Theory.

The arguments leading to the Special Theory, and the conclusions that follow from it, have an important bearing on our view of scientific reasoning and our apprehension of reality. These are the topics of Chapters Nine and Ten. Empiricist philosophies of science which have been current in the twentieth century and which are thought to obtain support from Einstein's theories of relativity are shown to be wrong. Although there is an important truth in empiricism, which must not be lost sight of, scientific argument is much more "rationalist" than philosophers have been willing to allow, and their reasoning is different in structure from the models usually proposed. Our apprehension of reality is therefore also different. Contrary to the claims of the Logical Positivists, we can meaningfully talk about things we cannot immediately verify, and although the Special Theory shows that some concepts are relative to certain frames of reference, it does not show that "everything is relative" but on the contrary establishes that other concepts are "absolute" in the relevant respects; and that although scientific knowledge is always to some extent tentative and liable to revision, reality is not inherently unknowable, but is something we can, cautiously and without arrogance, seek to know.

There is no isolated truth.

Millet

All things near and far
Hiddenly linked are.
Thou canst not stir a flower
Without the troubling of a star.

Blake

Contents

Chapter 1
The Integration of Space with Time

§1.1 Space and Time

This section develops the argument of *Space, Time and Causality*, ch.11

The great conceptual difference between the Special Theory and Newtonian mechanics is that whereas in the latter space and time are completely independent of each other, in the former they are integrated together into a single "spacetime". Minkowski, in a lecture in 1908, declared that henceforth space by itself, and time by itself, were doomed to fade away into mere shadows and only a kind of union of the two would preserve an independent reality. If that were self-evidently so, and we had an intuitive grasp of the unity of space and time, and a natural trade-off between distance and duration, it would be easy, with the aid of only a few unexceptionable extra assumptions, to argue for the Special Theory.[1]

But the assertion seems absurd: we go on talking about space and time in much the same way as we always did, with there still being an essential difference between them, namely that we can move in any direction and at different velocities in space, but only uniformly and inexorably forwards in time. The integration of space and time into spacetime has not abolished the distinction between the spatial and the temporal, but only the claim that they are entirely independent. But if that is all that is being claimed, we might think that it was not anything new, certainly not something to distinguish the Special Theory from the Newtonian mechanics that preceded it. The integration of space with time is something more than what we are familiar with in Newtonian physics where we already use four co-ordinates (x, y, z, t), to give the spatiotemporal position of an event, and the position of a particle will often depend on the time. The union of space and time in Newtonian physics is only a shot-gun marriage of convenience, and we can, whenever we want, separate them out into space and time again. It is only a notational union, the "Cartesian Product", $\Re^3 \times \Re^1$,

---

[1] See below, §5.3, pp.162-163.

of a three-space with a one-space. In the Special Theory, however, although the temporal dimension is still *different* from the spatial dimensions, it is, in a sense to be elucidated, not *separate* from it.

That there should be some intimate connexion between space and time should not surprise us, although in its time it was seen as a great surprise. The integration of space with time was really required by Newtonian mechanics, and the Special Theory can be seen in one way as the culmination of Newtonian principles. Although in its historical development the Special Theory was seen as being forced upon physicists as the only way to accommodate the facts of experimental observation, we can, also and more illuminatingly, see it as something to be aimed for in any case, and which, as it happens, yields electromagnetism as a corollary.

Some integration of space with time is suggested even within Newtonian mechanics itself. There is, in the first place, a suggestive parallel between angles and velocities: both a re-orientation and a uniform velocity leave Newtonian laws unaltered, and both an angular velocity and an acceleration (that is, the first derivatives of reorientation and velocity with respect to time) are associated with forces.[2] Much more fundamental is the principle of spatiotemporal continuity,[3] or locality as it is often called, which leads us to postulate some finite speed[4] as the maximum for the movement of material bodies or the propagation of causal influence, and so to constitute a fundamental trade-off between space and time.

---

[2] In defending Newton's concept of absolute space against Leibniz' criticisms, Clarke adapts Newton's bucket argument, which shows that angular velocities are associated with forces, and argues instead that accelerations are (Fourth Reply to Leibniz, §13, H.G.Alexander, ed., *The Leibniz-Clarke Correspondence*, Manchester, 1956, p.48; compare Isaac Newton, *Principia*, Scholium to Definition VIII, reprinted in Alexander, pp.157-160. See also J.R.Lucas, *A Treatise on Time and Space,* London, 1973, §47, pp.237-238.

[3] Mathematicians distinguish continuity from differentiability, but such refinements are ignored here, and when we speak of continuous manifolds, we shall assume that they are differentiable too.

[4] The words 'speed' and 'velocity' are often used interchangeably, but it is often convenient (though sometimes unacceptably pedantic) to keep the word 'speed' for the scalar magnitude, 186,000 miles per second, and the word 'velocity' for the vector, in which the direction is also specified.

We are led to locality by four considerations. Spatial continuity is characteristic of fields, and as we come to formulate Newtonian mechanics in terms of fields rather than forces we begin to appreciate the importance of spatial continuity in our view of the world. Differential equations likewise emphasize the importance of continuity, especially temporal continuity, and the laws of nature can best be expressed in terms of differential equations, then causal influence will be propagated in a spatiotemporally continuous way. It is, thirdly, an ideal of explanation. In particular, it is a concomitant of the programme of banishing "action at a distance" and developing instead the concept of a field. Although we are prepared to posit a causal connexion on the basis of repeated and carefully monitored concomitances, we look for a spatiotemporally continuous chain, or "cord", of propagation of causal influence from the cause to the effect.

Locality, fourthly, provides a criterion of identity, which otherwise would be hard to come by in the austere scheme of Newtonian metaphysics. The fundamental entities—point-particles or corpuscles or atoms—are alike in all respects, and distinguished from one another only in that two cannot be in the same place at the same time. That is enough to tell that two are different, but how do we tell whether or not it is one and the same point-particle occupying different places at different times? If it traversed a spatiotemporally continuous path between different positions at different times, it is the same entity: otherwise it is different.[5]

The argument from continuity to finitude is persuasive but not incontrovertible. Obviously if spatiotemporal is to be our criterion of identity, bodies must move with only finite speeds For an infinite speed of bodily motion would imply instantaneous, and hence discontinuous, movement to another place. There would be no reason why one thing which had ceased to exist in one place should be identified with one thing popping up at one place rather than another thing popping up at the same time somewhere else; or another thing popping up at some other time. So if we are to secure continuity, we must ban infinite speeds of motion of material body, and confine ourselves to finite speeds.

[5] For further discussion of the principle of locality, see below, §10.5. See also, J.R.Lucas, *A Treatise on Time and Space*, London, 1973, §§25-28, pp.118-134.

Equally, if gravitation, or any other causal influence, were propagated with an infinite velocity, so that its effects, even a long way off, were instantaneous with the cause, then there would be no reason why a particular cause should be acting instantaneously at one place rather than another.

But these arguments are not watertight. We might be able, under favourable conditions, to identify a particle popping up in one place with another that had dropped out of existence somewhere else immediately before. If we had an absolutely rigid body or an absolutely incompressible medium, an impulse would be transmitted through it instantaneously. If we had some impermeable shield and knew that causal influence was propagated only in straight lines, we could determine the causes of instantaneous effects by altering such shields.

All this is true, but not conclusive. Although infinite speeds would not lead to a straight self-contradiction, the conditions under which we could identify causal connexions or re-identify material bodies would be awkward. Long before the advent of the Special Theory absolutely rigid, inelastic bodies were an embarrassment to Newtonian physics. What happens when an inelastic fly is hit by an inelastic steam engine? An infinite force would be required to change the fly's velocity instantaneously to that of the engine. An absolutely rigid rod would require infinite forces throughout its length to ensure that every part acquired instantaneously the motion of an endpoint in collision with another rigid rod. Although a rigid rod would explain why the effect occurred where it did, it would not explain how it could come about, without invoking infinite forces, which a Newtonian is rationally reluctant to do.

It still is not a logically necessary consequence of this that there must be some maximum velocity. It could be the case that though no material particle moved, and no causal influence was propagated, infinitely fast, there was no upper bound to their speed of movement or propagation. But, again, it would be messy.[6] A physics in which there is a limit to which things tend that is itself incompatible with its guiding principles is a physics that is teetering on the brink of incoherence. If we accept the principle of

---

[6] For messiness, see §10.2; see also Graham Nerlich, "Special Relativity is not Based on Causality", *British Journal for the Philosophy of Science*, **33**, 1982, §1, p.363.

locality and seek a coherent physics, we are committed to there being some definite finite maximum velocity, which will indicate some profound trade-off between spatial and temporal intervals.

The argument of this section is carried further in §§5.3, and 10.5

§1.2 Implications of a Universal Speed

Many thinkers take it to be contrary to reason that there should be a finite maximum speed.[7] It seems to land us in paradox. If a photon is approaching with the speed of light, and we move towards it, it would seem that its speed relative to us must be increased, unless it is already infinite. But that is to assume that speeds in the same direction are simply additive, and there is no necessary reason why this should be so. Not every quantity need be additive. Consider gradients. If we have a gradient of 1 in 6, and another of 1 in 4 we cannot add them to get a gradient of 2 in 10, or of 5/12 $(= 1/6 + 1/4)$, or of any other simple combination. Instead, we have to go from gradients to angles, add the angles, and go from the resulting angle back to gradients. Angles are additive, whereas gradients are not. And similarly, one way of looking at the constancy of the speed of light is that it shows us that speeds in any given direction, like gradients, are not really additive, even though they may approximate to it at low speeds, and that we must look elsewhere for additivity where moving entities are concerned. It might then be thought that we had no warrant for supposing that if speeds are not additive, anything else would be. But it turns out that granted certain general conditions on the nature of the combination rule, all particular combination rules are members of the same family and can be transformed into one another by a "regraduation function". That is to say, there is some function which transforms the magnitudes that combine according to one combination rule into other magnitudes that combine according to some other combination rule; and in particular, we can always find

---

[7] Among them Descartes: see his letter to Beeckman, August 22 1634; reprinted in Charles Adam & Paul Tannery, *Oeuvres de Descartes*, Paris 1897, I, pp. 307–308; quoted by Gerald Holton, *An Introduction to Concepts and Theories in Physical Science*, 2nd ed., Reading, Mass., 1973, pp. 385–386, and by Wesley C.Salmon,"The Philosophical Significance of the One-Way Speed of Light", *Noûs*, **11**, 1977, p.254.

a function which transforms them into magnitudes that are additive. The conditions are that there should be a unit, or identity, element and a universal element, and that combination, which we shall indicate by the sign $\oplus$, should be associative, differentiable and, except for the identity and universal elements, strictly monotonic (that is, if the value of either of the variables is increased, then so also is the value of the resultant).

These five conditions can be expressed symbolically:

(1) there is an identity element O, such that $u \oplus O = u$, and $O \oplus u = u$;

(2) there is a universal element, in our case c, such that $u \oplus c = c$ and $c \oplus u = c$;

(3) for all u, v, w, the associative rule is satisfied

$$u \oplus (v \oplus w) = (u \oplus v) \oplus w; \qquad (1.2.1)$$

(4) the differentials $\mathrm{d}(u \oplus v)/\mathrm{d}u$ and $\mathrm{d}(u \oplus v)/\mathrm{d}v$ should exist and be continuous in u and v.

(5) $\mathrm{d}(u \oplus v)/\mathrm{d}u > 0$ and $\mathrm{d}(u \oplus v)/\mathrm{d}v > 0$, provided $u \neq 0$, $v \neq 0$, $u \neq c$, $v \neq c$.

Granted these five conditions, we can always construct a differentiable and strictly monotonic "regraduation" function $f(u)$, such that

$$f(u \oplus v) = f(u) + f(v), \qquad (1.2.2)$$

thus enabling us to re-measure speeds in such a way as to have simple addition as the combination rule.[8] Of course, these then will not be speeds in our sense or measured in our way, as distance covered divided by time. They will have been regraduated to be

---

[8] The proof, which is simple but tedious, is due to G.J. Whitrow, *Quarterly Journal of Mathematics*, 6, 1935, §4, pp. 252-6. Abel was the first to devise a regraduating function; among others who have devised them, often requiring slightly different conditions from the five listed here, are E.A. Milne, A.G. Walker, W.H. McCrea, J. Aczel and I.J. Good.

some other sort of magnitude—"rapidities" we might call them.[9]
Rapidities can be added. Zero rapidity corresponds to zero speed,
and infinite rapidity to the universal speed c. Conversely, we can
use the inverse of the regraduating function to define $\oplus$.

$$u \oplus v = f^{-1}[f(u) + f(v)]. \tag{1.2.3}$$

Thus provided there is a finite maximum speed c, there must be
some combination rule for speeds which will accommodate its being
a universal speed, and which, compounded with any other, results
in just itself.[10]

The standard combination rule given in the Special Theory is in
fact

$$u \oplus v = \frac{u + v}{1 + uv/c^2}, \tag{1.2.4}$$

which differs from simple addition only by the denominator itself
differing from unity by a very small amount for small values of u
and v. It is easy to check that

$$u \oplus 0 = u, \tag{1.2.5}$$

so the zero is the same in both cases. Similarly for the universal
speed, c,

$$u \oplus c = \frac{u + c}{1 + uc/c^2} = c. \tag{1.2.6}$$

This combination rule is easily shown to be associative, and is
clearly differentiable and monotonic. So it is a possible member
of the family which could be obtained from the simple addition
rule by a regraduating function. It is clearly a simple one. Others
no doubt could be devised, but would be less simple, which would

---

[9] J.M. Lévy-Leblond, "One more derivation of the Lorentz Transformation', *American Journal of Physics*, **44**, 1976, p.271; J.M. Lévy-Leblond
and J.M.Provost, "Additivity, Rapidity, Relativity', *American Journal of
Physics*, **47**, 1976, p.1045.

[10] The converse implication, from a universal to a maximum speed, need not
hold; see §3.7, p.113 below. The shift of emphasis from a maximum to
a universal speed may carry with it considerable consequences; see §10.5,
p.288.

both run counter to other requirements we are led to expect of a rule for compounding speeds, and would lead to awkwardnesses in the regraduating function.

Whittaker gives a direct argument for the combination rule which is simple and intuitive, although requiring more conditions on the concept of velocity than we have so far justified.[11] Any combination rule,

$$k = u \oplus v, \qquad (1.2.7)$$

is of the general form $g(u, v, k) = 0$, which must be symmetrical in u, v, and $-k$, since it can be viewed as a function of the relative velocities of three frames of reference A with respect to B, B with respect to C and (hence requiring a change of sign for k) C with respect to A. So, if we replace k by $-w$ the combination rule will be of the general form

$$g(u, v, w) = 0. \qquad (1.2.8)$$

This function, $g(u, v, w)$, Whittaker argues, must be not only symmetrical, but also linear, in u, v, and w.[12] So it must be of the form

$$l + m(u + v + w) + n(vw + wu + uv) + puvw = 0. \qquad (1.2.9)$$

where l, m, n and p are constants to be determined. When u and v are both zero, w must be zero too: hence $l = 0$. More generally, if $v = 0$, $w = -u$. So $l + n(-u^2) = 0$, for all values of u; so, since $l = 0$, $n = 0$.

From the universal property, if $v = c$, $w = -c$. So

$$mu - puc^2 = 0. \qquad (1.2.10)$$

Hence $m = pc^2$ and, replacing w by $-k$, we obtain the combination rule

$$k = \frac{u + v}{1 + uv/c^2}. \qquad (1.2.11)$$

We are not, as yet, in a position to justify Whittaker's symmetry condition, nor is his argument for linearity absolutely watertight.[13]

---

[11] Sir Edmund Whittaker, *From Euclid To Eddington*, Cambridge, 1948, reprinted Dover, 1958, §21, pp.49-51.

[12] For otherwise, if, for example, it were quadratic in u, there would be two values of u which would, when combined with v, produce the same w. For the same reason it must be linear in v, and, again, linear in w.

[13] See further below, §5.2, pp.160-162.

Nevertheless, his derivation is persuasive, and we may explore the consequences of adopting that combination rule, as being the simplest one that satisfies the symmetry and other requirements it is natural to impose.

The formula

$$u \oplus v = \frac{u + v}{1 + uv/c^2} \qquad (1.2.12)$$

is rather like the formula, familiar from school trigonometry, for $\tan(u + v)$, namely

$$\tan(u + v) = \frac{\tan u + \tan v}{1 - \tan u \tan v}. \qquad (1.2.13)$$

It is even more like the formula for the "hyperbolic tangent"

$$\tanh(u + v) = \frac{\tanh u + \tanh v}{1 + \tanh u \tanh v}. \qquad (1.2.14)$$

This suggests regraduating from speeds to rapidities by means of the regraduating function

$$u = c \tanh \phi. \qquad (1.2.15)$$

A more formal derivation, due to Ramakrishnan,[14] is obtained by setting $v = -u - du$ and $k = -cd\phi$.

Then $\quad - cd\phi(1 - u(u + du)/c^2) = -du,$

$$i.e. \quad cd\phi = \frac{du}{1 - u^2/c^2}. \qquad (1.2.16)$$

Integrating, $\quad u = c \tanh \phi.$

Thus the mapping of the real line $-\infty < \phi < +\infty$ onto the finite segment $-c < u < +c$ is given by $u = c \tanh \phi$

The regraduating function

$$\phi = \tanh^{-1}\left(\frac{u}{c}\right) \qquad (1.2.17)$$

---

[14] Given by O.Costa de Beauregard, *Foundations of Physics*, **16**, 1985, pp.1154–5.

has the properties of being continuous, differentiable and strictly monotonic, of being 0 when $u = 0$, and of tending to infinity as u tends towards c, and to minus infinity as u tends towards $-c$. Although not unique in these properties, $\tanh^{-1}$ is simpler than any other function possessing them. Any simple function, quadratic, quartic, etc., will fail to be strictly monotonic, as will most cubic, quintic, etc., functions. For the same reason trigonometric functions are unsuitable. Linear functions take infinities into infinities. Logarithmic and exponential functions remain, and of these $\tanh^{-1}(u/c)$, which is $\frac{1}{2}(\log(1 + u/c) - \log(1 - u/c))$, and $\tanh r$, which is $(e^r - e^{-r})/(e^r + e^{-r})$, are the simplest. So the easiest way of accommodating the fact of there being a finite universal speed, c, to speeds otherwise seeming to obey a simple addition rule, is to regraduate them by considering each speed as a fraction of the universal speed, and taking the inverse hyperbolic tangent. In this way we can convert from speeds into rapidities, which are simply additive.

Thus if we regraduate from speeds to rapidities, we need to use the inverse hyperbolic tangent. Or, looking at it the other way round, speeds are hyperbolic tangents of rapidities as gradients are ordinary tangents of angles. A slope of 30° has a gradient of 1 unit up for $\sqrt{3}$ units along the horizontal, and one of 45° has a gradient of 1 unit up for 1 unit along. Just as we cannot add gradients directly in the way we can add angles, so we cannot add speeds directly in the way we can rapidities. As we saw, if we want to compound the gradients 1 in $\sqrt{3}$ and 1 in 1, the answer will not be 2 in ($\sqrt{3} + 1$): rather, first we must take the inverse tangent to obtain the corresponding angles, then add the angles, and then take the tangents of the sum. Thus, according to this rule of combination,

$$\frac{1}{\sqrt{3}} \oplus \frac{1}{1} = \tan(\tan^{-1}\frac{1}{\sqrt{3}} + \tan^{-1}\frac{1}{1}) = \tan(30° + 45°)$$

$$= \tan 75°. \tag{1.2.18}$$

The standard trigonometric rule for $\tan(x + y)$, *viz*

$$\tan(x + y) = \frac{\tan x + \tan y}{1 - \tan x \tan y}, \tag{1.2.19}$$

offers a very close parallel to our rule, differing only in the minus sign in the denominator. That is, for gradients

$$x \oplus y = \frac{x + y}{1 - xy},$$
(1.2.20)

whereas for speeds

$$u \oplus v = \frac{u + v}{1 + uv}.$$
(1.2.21)

We have thus shown that if there is a universal speed, and if velocities in the same direction are compounded according to a rule that is associative, differentiable, monotonic, and subject to natural symmetry condition, then the combination rule is like that for hyperbolic tangents, and the inverse hyperbolic tangent is the required regraduating function.

The argument of this section is developed further in §1.6

§1.3   Pseudo-Trigonometry

This section is primarily expository.
The main argument continues in §1.4

Thus far the analogy between hyperbolic and ordinary trigonometry is only illustrative. In fact it is much deeper than that, and so is worth considering more closely. The comparison between speeds and gradients can be brought out by re-naming rapidities "pseudo-angles". Whereas directions in space, or the space-like subspace of spacetime, subtend angles, "directions" in spacetime which are not entirely space-like but partially time-like, subtend pseudo-angles. Two things moving with a uniform speed are "oriented" to each other at a pseudo-angle as two things facing different ways are oriented at an ordinary angle to each other. Not only is speed like orientation, but the first derivative of velocity, acceleration, is like the first derivative of orientation, angular velocity. The fact that the latter two are associated with forces in Newtonian mechanics, while the former two are not, no longer seems surprising, but is a manifestation of the way in which time is somewhat like space, only giving rise to hyperbolic pseudo-angles instead of ordinary circular angles that occur in geometry and trigonometry.

Hyperbolic functions are defined in terms of exponentials, and their inverses in terms of natural logarithms, and are very similar to the ordinary trigonometrical functions, except that their exponents differ by a factor i, the square root of minus one; cos and cosh, in particular, are very similar, the one having imaginary, the other real, exponents. If there is a universal speed, the combination rule must involve a hyperbolic function, and so will differ from ordinary trigonometric functions only in there being a factor i at various places in the expression. We already have a trade-off factor c. If we want to apply comparable units to time and space—as for example when we measure distance in "light-years"—we multiply time by the speed of light. Instead of using co-ordinates (x, y, z, t) with x, y and z, measured in conventional units of distance—miles or metres—and in conventional units of time—seconds or years—we should multiply t by c, so as to measure it in the same units; instead of talking of one second, we should call it 186,000 miles or 300,000,000 metres. So we use (x, y, z, ct). If we now add in a further factor i, and, to avoid confusion, have $x_1 = x$, $x_2 = y$, $x_3 = z$, and $x_4 = ict$, we have a simple Euclidean four-dimensional space which when translated back into (x, y, z, t) yields the pseudo-angles and hyperbolic functions we were forced to introduce in order to accommodate the finite universal speed c. The time-like dimension in spacetime is still different from the space-like dimensions: but the way in which it differs is like that in which ict differs from x.

The hyperbolic functions are related among themselves in much the same way as the ordinary trigonometric ones, only with changes of sign. Thus instead of the familiar $\cos^2 \theta + \sin^2 \theta = 1$, we have $\cosh^2 \theta - \sinh^2 \theta = 1$. And just as we can define cos and sin in terms of tan, so we can define cosh and sinh in terms of tanh, and the hyperbolic tangent of a pseudo-angle, that is, a rapidity, is a velocity. More precisely, we have

$$v = c \tanh \phi$$

$$\text{So that } \cosh \phi = \frac{1}{\sqrt{1 - v^2/c^2}} \tag{1.3.1}$$

$$\text{and } \sinh \phi = \frac{\tanh \phi}{\sqrt{1 - \tanh^2 \phi}} = \frac{v/c}{\sqrt{1 - v^2/c^2}}.$$

Since these formulae keep on turning up, it is convenient to simplify them. We abbreviate $\tanh\phi$ as β, and $\cosh\phi$ as γ.[15] Since we shall often have occasion to use them, we give their definitions explicitly:

$$\beta = \frac{v}{c} \quad \text{and} \quad \gamma = \frac{1}{\sqrt{1 - v^2/c^2}} \equiv (1 - (v^2/c^2))^{-\frac{1}{2}}. \quad (1.3.2)$$

The analogy between velocity and pseudo-angles gives a key to understanding the Lorentz transformation. The Lorentz transformation, of which the more familiar Galilean transformation can be seen as a special case,[16] is the key to understanding the Special Theory, but is hard at first to comprehend. What otherwise seems a deeply obscure set of equations becomes easy to grasp once we view a change from one frame of reference to another moving at a uniform translational[17] velocity with respect to it as like a change from set of co-ordinates to another set at an angle to it.

§1.4 The Lorentz Transformation

The Lorentz transformation can be seen as an immediate consequence of the programme of integrating space with time, although we shall obtain it in a less sophisticated way in the next chapter.[18] It connects the spatial and temporal co-ordinates of a point in one frame of reference with those in another frame of reference moving relative to it with a uniform velocity. Without loss of generality we can take the direction of relative motion along the X-axis. If we now consider the four-dimensional analogue of our familiar three-dimensional Euclidean space in which we represent points by the

---

[15] Einstein himself used β to abbreviate $(1 - v^2/c^2)^{-\frac{1}{2}}$, but now γ has become more usual.

[16] See more fully below, §2.6.

[17] In his original paper ("Zur Electrodynamik bewegter Körper", *Annalen der Physik*, **17**, 1905, p.891) Einstein used the word 'translational' to distinguish motion in a straight line from angular velocity, or rotation, under which the laws of physics are not covariant. We shall follow the current practice of omitting the word 'translational', and assume that a velocity is translational, unless it is stated to be angular.

[18] See below, §2.7. See also, more fully, ch.5.

co-ordinates (x_1, x_2, x_3, x_4), where $x_1 = x$, $x_2 = y$, $x_3 = z$ and $x_4 = ict$, then a rotation in the (x_1, x_4) plane through an angle θ transforms the point (x_1, x_2, x_3, x_4) to the point (x_1', x_2', x_3', x_4') where

$$
\begin{aligned}
x_1' &= x_1 \cos \theta + x_4 \sin \theta \\
x_2' &= x_2 \\
x_3' &= x_3 \\
x_4' &= - x_1 \sin \theta + x_4 \cos \theta
\end{aligned}
\tag{1.4.1}
$$

Since the second and third components are unchanged, we shall nearly always leave them out.

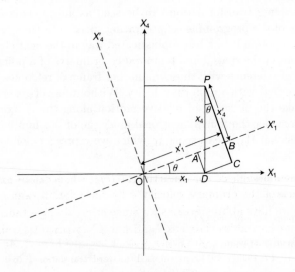

Figure 1.4.1 In ordinary Euclidean space, if we turn the axes through θ, then x_1', the co-ordinate of the point P along the X_1' axis, will be $OB = OA + AB = x_1 \cos \theta + x_4 \sin \theta$; and x_4', the co-ordinate of the point P along the X_4' axis will be $PB = PC - CB = PC - AD = x_4 \cos \theta - x_1 \sin \theta$.

We can write these equations in matrix notation[19]

$$\begin{pmatrix} x_1' \\ x_4' \end{pmatrix} = \begin{pmatrix} \cos\theta & \sin\theta \\ -\sin\theta & \cos\theta \end{pmatrix} \begin{pmatrix} x_1 \\ x_4 \end{pmatrix}. \qquad (1.4.2)$$

If we now go back to the (x, y, z, ict) notation we have

$$x' = x\cos\theta + ict\sin\theta, \quad \text{and} \quad ict' = x\sin\theta + ict\cos\theta; \qquad (1.4.3)$$

the latter can be rewritten

$$t' = -\frac{x\sin\theta}{ic} + t\cos\theta \qquad (1.4.4)$$

Since, as we have noted, hyperbolic and ordinary trigonometrical functions are very similar, differing chiefly, when expressed in terms of exponentials, in the exponents being real instead of imaginary, we can replace θ by $i\phi$. Then, as we saw before, $\cos i\phi = \cosh\phi$ and $\sin i\phi = i\sinh\phi = -\frac{1}{i}\sinh\phi$. So our transformation becomes

$$x' = x\cosh\phi - ct\sinh\phi \qquad (1.4.5)$$

$$t' = -\frac{x}{c}\sinh\phi + t\cosh\phi, \qquad (1.4.6)$$

or in the matrix notation,

$$\begin{pmatrix} x' \\ t' \end{pmatrix} = \begin{pmatrix} \cosh\phi & -c\sinh\phi \\ -\frac{1}{c}\sinh\phi & \cosh\phi \end{pmatrix} \begin{pmatrix} x \\ t \end{pmatrix}. \qquad (1.4.7)$$

The pseudo-angle ϕ represents the rapidity, which can be expressed in terms of the ordinary velocity v by the relation $\tanh\phi = v/c$, which we have abbreviated as β. Since $\cosh\phi = \{1 - \tanh^2\phi\}^{-\frac{1}{2}}$, which we have abbreviated as γ and $\sinh\phi = \tanh\phi\{1 - \tanh^2\phi\}^{-\frac{1}{2}}$ the Lorentz transformation becomes

$$x' = \gamma(x - vt)t' = \gamma(\frac{-\beta x}{c} + t). \qquad (1.4.8)$$

---

[19] An introduction to matrix notation may be found in any standard textbook, such as A.E. Coulson, *An Introduction to Matrices*, Longmans, 1965.

or, written out in full,

$$x' = \frac{x}{\sqrt{1-(v/c)^2}} - \frac{vt}{\sqrt{1-(v/c)^2}}$$

$$t' = \frac{-xv/c^2}{\sqrt{1-(v/c)^2}} + \frac{t}{\sqrt{1-(v/c)^2}}, \tag{1.4.9}$$

or, in the matrix form,

$$\begin{pmatrix} x' \\ t' \end{pmatrix} = \begin{pmatrix} \frac{1}{\sqrt{1-v^2/c^2}} & \frac{-v}{\sqrt{1-v^2/c^2}} \\ -\frac{v/c^2}{\sqrt{1-v^2/c^2}} & \frac{1}{\sqrt{1-v^2/c^2}} \end{pmatrix} \begin{pmatrix} x \\ t \end{pmatrix}$$

$$= \begin{pmatrix} \gamma & -v\gamma \\ -v\gamma/c^2 & \gamma \end{pmatrix} \begin{pmatrix} x \\ t \end{pmatrix}. \tag{1.4.10}$$

In the *ict* notation this takes the simple form

$$\begin{pmatrix} x' \\ ict' \end{pmatrix} = \begin{pmatrix} \gamma & i\beta\gamma \\ -i\beta\gamma & \gamma \end{pmatrix} \begin{pmatrix} x \\ ict \end{pmatrix}. \tag{1.4.11}$$

The Lorentz transformation has its counter-intuitive aspects, most notably in having t' depend on x: these will be discussed in the next chapter. For the present we continue the project of integrating space with time, and consider its bearing on the other basic concepts of spacetime physics.

A conceptually simpler derivation of the Lorentz
transformation is given in §2.7, and others in ch.5

§1.5 Spacetime Concepts

The fact that there was a universal speed led us to use hyperbolic functions, and to view a uniform velocity as a pseudo-angle. This in turn leads to a fundamental shift in our concepts of space and time, and the corresponding physical concepts of momentum and energy. Spatial and temporal separation are no longer seen as fundamental, but as "perspectival" aspects of a spacetime separation, which is regarded as the really fundamental concept.

Under a rotation in ordinary three-dimensional Euclidean space the X-, Y-, and Z-co-ordinates are altered, but lengths and distances remain the same, and similarly under a "rotation" through a pseudo-angle in Minkowski spacetime the spacetime analogue of distance should again remain the same. In a Euclidean four-dimensional space, the four-dimensional analogue of Pythagoras' theorem gives the distance between two "points" $(x_1^*, x_2^*, x_3^*, x_4^*)$ and $(x_1^\dagger, x_2^\dagger, x_3^\dagger, x_4^\dagger)$ as

$$\sqrt{(x_1^* - x_1^\dagger)^2 + (x_2^* - x_2^\dagger)^2 + (x_3^* - x_3^\dagger)^2 + (x_4^* - x_4^\dagger)^2} \qquad (1.5.1)$$

or,

$$\{(x_1^* - x_1^\dagger)^2 + (x_2^* - x_2^\dagger)^2 + (x_3^* - x_3^\dagger)^2 + (x_4^* - x_4^\dagger)^2\}^{1/2} \qquad (1.5.2)$$

Translated into (x, y, z, ict) terms, the distance between two events, (x^*, y^*, z^*, t^*) and $(x^\dagger, y^\dagger, z^\dagger, t^\dagger)$ becomes

$$\{\{(x^* - x^\dagger)^2 + (y^* - y^\dagger)^2 + (z^* - z^\dagger)^2 - (ct^* - ct^\dagger)^2\}^{1/2} \qquad (1.5.3)$$

This is the **spacetime separation**. It is invariant under the Lorentz transformation, and is the fundamental spatiotemporal concept of the Special Theory.

Spacetime separation differs from ordinary spatial distance in Euclidean space in not being "positive definite". In Euclidean space, with the Pythagorean rule

$$d = \sqrt{(x^* - x^\dagger)^2 + (y^* - y^\dagger)^2 + (z^* - z^\dagger)^2}, \qquad (1.5.4)$$

we can be sure that the term under the square root will never be negative, and will always be positive unless the two points are coincident. With the modified rule we can no longer be sure. The term under the square root,

$$(x^* - x^\dagger)^2 + (y^* - y^\dagger)^2 + (z^* - z^\dagger)^2$$

will be positive or negative, and therefore the spacetime separation itself real or imaginary, according as to whether the difference between the space-like intervals is greater or less than that between the time-like ones. If we think of spacetime separation as a generalisation of spatial distance, it fits the way imaginary numbers were introduced to have it real when the space-like component was larger than the time-like one, and in that case we denote it by σ, and write

$$\sigma = \sqrt{(x^* - x^\dagger)^2 + (y^* - y^\dagger)^2 + (z^* - z^\dagger)^2 - (ct^* - ct^\dagger)^2} \quad (1.5.5)$$

and say that the spacetime interval is "space-like" when σ is real, "light-like" when σ is zero, and "time-like" when σ is imaginary.[20] But space-like separations are less important than time-like separations, because different events in the history of a material object must have a time-like separation, and an event cannot cause, or be caused by, another event that has a space-like separation from it.[21] So it is sensible to have the signs of the spatial and temporal separations the other way round. In that case we use the Greek letter, τ, where $\tau^2 = -\sigma^2/c^2$, or, $\sigma = ic\tau$, or explicitly

$$\tau = \sqrt{\left((t^* - t^\dagger)^2 - \left(\frac{x^* - x^\dagger}{c}\right)^2 - \left(\frac{y^* - y^\dagger}{c}\right)^2 - \left(\frac{z^* - z^\dagger}{c}\right)^2 \right)}$$

$$(1.5.6)$$

Where τ is measured along the path of some material object moving from one location in space at one time to another at another, it is said to be the "proper time" *of* that object, or, more precisely the proper time that has elapsed for that object between the two events.[22] It is customary to indicate the convention about signs

---

[20] See further below, §2.2.

[21] See below, §2.2.

[22] See further below, §2.12.

being used by means of the "Lorentz signature" of spacetime. If the space-like dimensions are assigned a positive value and t he time-like one a negative in the pseudo-Pythagorean rule for spacetime separation, the Lorentz signature is $(+++-)$. Except in Chapter 5,[23] we shall follow the opposite convention, and have the Lorentz signature $(+---)$, that is have the space time separation between two events, (x^*, y^*, z^*, t^*) and $(x^\dagger, y^\dagger, z^\dagger, t^\dagger)$ given by 1.5.6 above.

Ordinary spatial and temporal intervals can be defined in terms of proper time, with the aid of pseudo-trigonometry. Thus if the proper time is τ, and the pseudo-angle or rapidity ϕ,

$$(t - t^\dagger) = \tau \cosh \phi. \qquad (1.5.7)$$

Similarly, using $\cosh^2 \phi - \sinh^2 \phi = 1$,

$$\{(x - x^\dagger)^2 + (y - y^\dagger)^2 + (z - z^\dagger)^2\}^{1/2} = c\tau \sinh \phi \qquad (1.5.8)$$

What we have done essentially is to define magnitudes, τ which, in virtue of the way $\tanh \phi$ was defined, enables us to express, in terms of τ and the pseudo-angle ϕ, the temporal interval and spatial distance between two events, that is to say the time taken and the distance covered between departure and arrival, and hence also the speed, as ordinarily understood.

But we need to be careful. In ordinary life there is slight unclarity about the concept of a gradient. Map-makers are concerned with distances along the horizontal, and mean by 1 in 3 one unit up for each 3 units along the horizontal; i.e. a slope of 30°. Similarly 1 in 1 is 45°. This is the usage adopted previously. But road-users are not so much interested in map-reading as the distance traversed by their wheels. For them a slope of 30° is 1 in 2, and 45° would be not 1 in 1 but 1 in $\sqrt{2}$. They use the sine of the angle rather than the tangent. For the gentle inclinations normally encountered in England the difference is negligible, but with larger angles it is not: whereas for the map-maker a 1 in 1 gradient is 45°, for the road user it is a vertical ascent. So too with the hyperbolic functions, the difference between $\tanh \phi$ and $\sinh \phi$ is negligible for small ϕ, but very significant as ϕ becomes large. Suppose a body is moving along the X-axis (so that $y = 0$ and $z = 0$ throughout). Then

---

[23] §5.3, pp.162ff.

$\mathrm{d}x/\mathrm{d}\tau = c\sinh\phi$, $\mathrm{d}t/\mathrm{d}\tau = \cosh\phi$ and therefore, as we might put it, in order to have the same measure for time as for space

$$\frac{\mathrm{d}x}{c\mathrm{d}t} = \tanh\phi \qquad (1.5.9)$$

The difference between $\tanh\phi$ and $\sinh\phi$, and hence between $\mathrm{d}x/c\mathrm{d}t$ and $\mathrm{d}x/\mathrm{d}\tau$ is very small for small ϕ. To the extent that we have good reason for believing τ more fundamental than x and t, and so $\mathrm{d}\tau$ than $\mathrm{d}x$ and $\mathrm{d}t$, we should think that instead of the velocity of classical space-and-time physics, $\mathrm{d}x/\mathrm{d}t$, (or $\mathrm{d}x/c\mathrm{d}t$) we ought in spacetime physics to reckon the way in which distance varies with spacetime separation in terms of $\mathrm{d}x/\mathrm{d}\tau$.

Other adjustments will be called for too. In particular, the space-time version of momentum, which in classical physics is mv, *i.e.* $m\mathrm{d}x/\mathrm{d}t$ along the X-axis, and correspondingly, $m\mathrm{d}y/\mathrm{d}t$ along the Y-axis, and $m\mathrm{d}z/\mathrm{d}t$ along the Z-axis will be $m\mathrm{d}x/\mathrm{d}\tau$, $m\mathrm{d}y/\mathrm{d}\tau$ and $m\mathrm{d}z/\mathrm{d}\tau$ respectively. If we make a similar adjustment for the fourth co-ordinate and consider $m\mathrm{d}t/\mathrm{d}\tau$. We have

$$m\frac{\mathrm{d}t}{\mathrm{d}\tau} = m\cosh\phi = \frac{m}{2}(e^{\phi} + e^{-\phi}) \qquad (1.5.10)$$

If ϕ is small, we can use the exponential expansion to obtain $\cosh\phi \approx 1 + \frac{1}{2}\phi^2$. Also for small ϕ

$$\phi = \tanh^{-1}\left(\frac{v}{c}\right) \approx \frac{v}{c} \qquad (1.5.11)$$

Thus the time analogue of momentum in a spatial direction is approximately $m(1 + \frac{1}{2}v^2/c^2)$ which is $m + \frac{1}{2}mv^2/c^2$. The second term is very like the familiar kinetic energy of Newtonian mechanics, with the velocity expressed as a fraction of the velocity of light. The first term is simply the mass. The time analogue of momentum is compounded of mass and energy—"massergy" as it is sometimes called —which can be expressed either as mass or as energy. If we measure velocity in ordinary units rather than as a fraction of the velocity of light, and are dealing with kinetic energy as ordinarily measured, we need to multiply both terms by c^2, to keep them comparable. Then (massergy $\times c^2$) will be

$$mc^2 + \frac{1}{2}mv^2 \qquad (1.5.12)$$

The second term being the kinetic energy as ordinarily measured and the first expressing Einstein's famous equivalence between mass and energy.

The concept of a pseudo-angle thus leads to a natural treatment of spacetime, in which angles between ordinary space-like directions are paralleled by pseudo-angles where time-like "directions" are involved, and where the Pythagorean rule for spatial distance is paralleled by the pseudo-Pythagorean rule for spacetime separation. It is clearly possible to express spatial and temporal intervals in terms of the spacetime separation and pseudo-angle, but it is not necessary to do so—we clearly can express the latter in terms of the former, and we could use, say, pseudo-angle and time as our basic concepts. It is plausible to hold that spacetime separation will play as fundamental a part in spacetime geometry as distance does in ordinary Euclidean geometry. If it is fundamental, then it is natural to generalise the three-vector concept of momentum in Newtonian mechanics to a four-vector concept in the Special Theory in which momentum is almost the same, being $m\mathrm{d}x/\mathrm{d}\tau$ etc. instead of $m\mathrm{d}x/\mathrm{d}\tau$ etc., and the fourth vector is one of massergy, which encompasses both the mass and the kinetic energy of Newtonian mechanics. The programme of integrating space and time thus leads to an altered perspective on the concepts of Newtonian mechanics, which are neither left untouched nor made completely different. Energy is seen as the time-like correlate of momentum, and both are defined with reference to spatiotemporal separation rather than spatial and temporal intervals as hitherto. If we are to integrate space and time, we must accept not only a new geometry of spacetime—Minkowski geometry—but a new physics as well, the physics of the Special Theory.

The argument of this section is further developed
in §§2.1, 2.2, 3.2, 5.3, and more fully in ch.6

§1.6 The Status of Spacetime

The classical concept of velocity adopted by Newtonian mechanics can now be seen as a special case in which the universal velocity is infinite, and the composition rule takes the peculiarly simple form of simple addition. It has the undoubted merit of extreme simplicity. It also fits both our intuitive sense that time is very unlike space, and Newton's metaphysical assumptions. Newton not only takes space and time as separate, fundamental categories, but introduces matter as a fundamental category too. Newtonian point-particles are idealised things, which exist throughout some interval of time and can change without losing their identity. Space provides room for change. Spatial position is what a point-particle, or corpuscle, or basic particular, must have, though no particular position is bespoken to any particular point-particle; a point particle must at any one time have some spatial position, but it can change its position. Space therefore provides, in the Newtonian system, the possibility of a point-particle occupying some position or other, without that position being necessarily occupied by anything. Hence the sense of empty space being a void non-entity, and therefore as featureless as possible, a mere receptacle, devoid of all properties of its own.[24] It is where things might be, but actually are not. It is an adjective, an adjective of potentiality, not a proper noun referring to something real. What is real in the corpuscularian philosophy is matter, idealised things. Space in itself is unreal, unless it can be seen as the *sensorium* of God in the way Newton saw it.[25]   And space is separate from time, being separated by matter. Time implies change, and change implies matter, since if there is change there must be something that persists through change, which changes and yet remains the same thing. It is only when we have matter changing that we need space as a receptacle for it to be in and to change in. The order is Time, Matter, Space, and Space is therefore both separate from Time, and not really a substance but only an attribute.

The Newtonian, or corpuscularian, scheme has great merits, metaphysical as well as scientific, but has the demerit of giving

[24] See further, Graham Nerlich, "What Can Geometry Explain?", *British Journal for the Philosophy of Science*, **30**, 1979, pp.69-83. See also, §5.2, pp.157-158, and §§8.3-8.5.

[25] See below, §8.6.

no fundamental trade-off between time and space, and unless the universal velocity be taken as being only a limit, never actually realised in fact, it has the further disadvantage of breaching the principle of continuity, required by Newtonian mechanics both as a criterion of identity for material bodies and as an ideal for causal connexion.[26] These two reasons, though neither on its own compelling, both point to there being a finite universal velocity, which then introduces the inverse hyperbolic tangent as the regraduating function, and suggests the analogy between angles and velocities, and between angular velocities and accelerations.

We can also argue from field theories in support of the programme of integrating space and time into a unified spacetime, with the consequent need for some trade-off between space and time. Field theories do not have to be spatiotemporal: gravitation can be expressed as a purely spatial field. But where fields change with time, it is natural to view them as spatiotemporal. In any case field theories alter our perspective on space. Instead of being a mere receptacle, where things happen to be, space seems more of an entity, a plenum, having a function defined from it into the real numbers (or into a set of vectors etc.). It seems more of a noun, referring to a something that has a field, instead of being a rather vacuous quasi-quality which tells us something about things, namely what position they have. Once this change of perspective is accomplished, space seems less different from time, and seems to be more of an entity, more likely to be a partial aspect of an even more real entity, such as spacetime. If space is a void non-entity, as it seems to be in the corpuscularian or Newtonian scheme, it is difficult to unify it with anything. The more substantial field theory makes it appear, the more ready it is to be integrated to form an even more substantial reality.

The shift from Time and Space to Spacetime thus represents a profound change in our fundamental categories of thought, a shift away from Newton, and before him Democritus and Epicurus, who all believed in atoms and the void, towards Descartes and Spinoza, and before them the Stoics, who reified space, and were precursors of modern field theories. In this respect the Special Theory is quite rightly contrasted with the Newtonian mechanics that preceded it. But Newtonian mechanics was already committed to some principle

[26] See above, §1.1, and below, §10.5.

of continuity, both as a criterion of identity for individual point-particles, and as a condition of causal explicability. In that respect, therefore, the Special Theory is to be seen as a continuation of Newtonian themes culminating in a theory of the propagation of causal influence (electromagnetism) together with re-identifiable point-particles.

The shift from Time and Space to Spacetime is not only profound, but difficult. It is easy to think of spacetime as being just like space, only with an extra dimension, and thus to spatialise time, and give an account of reality which is entirely static, in which there is no real change or process. Weyl puts it forcefully:

> The objective world simply *is*, it does not *happen*. Only to the gaze of my consciousness, crawling along the life line of my body, does a section of this world come to life as a fleeting image in space which continuously changes in time.[27]

This is a mistake. Time is not like space, and we do not crawl along world lines. But it is a very natural mistake, and much of the effort of this book is devoted to sharpening up our understanding of the concepts of the Special Theory, in order that we may not be unconsciously propelled towards fallacious arguments and false conclusions.

The argument of this section is further developed
in ch.6 and in §2.12, and §§8.5-8.7

---

[27] Hermann Weyl, *Philosophy of Mathematics and Natural Science*, 2nd ed. New York, 1963, p.116.

§1.7   The Michelson-Morley Experiment

In this chapter we have argued on the basis of a vague principle of locality or continuity that there ought to be a universal finite speed, implying some intimate relation between time and space. It might therefore be asked whether the result of the Michelson-Morley experiment could have been predicted. Some have maintained that it could,[28] but that is not how the history of the development of the Special Theory is told. It was, we are told, the null result of the Michelson-Morley experiment that forced physicists to acknowledge a finite universal speed in the theory of electromagnetism, and thus paved the way for the acceptance of the Special Theory. It was an experimental result, not some *a priori* reasoning, that led to new understanding. Furthermore, it was taken not just as an experimental result, but as a *brute* empirical fact, which forces us to adopt the Special Theory, as alone able to accommodate the data of observation. But this is both historically and philosophically false. Although the Michelson-Morley experiment was later adduced as strong empirical evidence in favour of Einstein's Special Theory, and although it is often said that Einstein developed his Special Theory in order to account for the null result of the Michelson-Morley experiment, as a matter of historical fact it had rather little influence on his actual thinking. Einstein first of all thought in a general way about the phenomena of nature, and sought "the beautiful simplicity" that, he was convinced, must lie beneath the tangled phenomena of experience. Only after he had worked out the mathematical implications of some new position would he search for experimental data which might test its validity.[29]

Philosophically, too, the null result, although undoubtedly an empirical fact, cannot be just a brute empirical fact, or it would not be susceptible of any explanation. But we have been able, though only after the event, to discern the rationale of the speed of light being a universal speed, and so by the same token have a

---

[28] See further below, §10.7.

[29] E.M.Mackinnon, *Scientific Explanation and Atomic Physics*, Chicago University Press, 1982; G.Holton, *Thematic Origins of Scientific Thought: Kepler to Einstein* Harvard University Press, 1973, pp.219-352. A further reference is given in §5.1, p.151.

line of argument which could have led us, had we been sufficiently intelligent, to anticipate, on purely rational grounds, the result of empirical experiment. It is, for many, an unwelcome conclusion. We are faced here, as often in physics and especially in the Special and General Theories, with a consequence of theoretical and empirical arguments which have not been adequately studied in the philosophy of science. The official philosophy of science is empiricist: it just happens to be a fact that the speed of light is the same in all frames of reference, and we explain this fact by positing some theory to account for it. But the argument of this chapter has shown that it is not just a fact, and as a matter of historical fact the Special Theory was not originally posited to account for it. Considerations of beauty, of simplicity, of coherence, of symmetry, of elegance, of rational explicability, all are appealed to in formulating a physical theory, and often in defending it against adverse evidence. If the empirical evidence is against a scientific theory, we do not just abandon the theory without more ado, but strive in all sorts of ways to explain away the evidence or adjust the theory so as to retain it in spite of the adverse evidence. Only if the evidence is repeated again and again and cannot be plausibly discounted at all, are we prepared to jettison a theory we are attracted to on other grounds.

We are thus led to reject radical empiricism as an adequate philosophy of science. It does not fit the facts. But what then should we put in its place? We cannot go to the other extreme, as Plato did, and the Rationalists—that is to say Descartes, Spinoza and Leibniz—are said to have done. Although the physicist is largely guided by his physical intuition, he acknowledges the need for empirical test somewhere along the line. But there is a lot of play between theory and experimental observation. If predictions are not confirmed, it may be that the observation was clumsily carried out; or it may be that it was competently carried out, but badly designed; or it may be that it was competently carried out and well designed, but the theory was not quite what we thought it was; or it may be that it was competently carried out, well-designed, within the range of applicability of the theory, but subject to some other, unforeseen, and unforeseeable, interference. These different ways of explaining untoward results are usually all lumped together, but that is a mistake. We need to consider them separately, because they illuminate the interplay between rationalist and empiricist arguments in arriving at scientific truth. In general, an experiment should be seen as a *questioning* of nature. Usually the scientist has two or more alternative answers in mind, and seeks to discover which is the right one. He may fall down in not taking care to use, so to speak, clear and precise language—if he weighs things on crude scales or in a draught, if he does not insulate electric wires, etc. This is simple experimental error, common among schoolboys but less so among professional experimentalists. More often the question is precisely, but wrongly, formulated. It may be in terms of the wrong concepts, or it may be that entirely the wrong conditions are being specified: vital force, calorific fluid and phlogiston were inadequate concepts; it is no good asking whether heavy bodies stay still or continue to move unless frictional and gravitational forces are specified. The experiments and observations which led to the adoption of Newtonian mechanics were both competently carried out and well-designed: their fault, in so far as it is one, was that the range of applicability of Newtonian mechanics was assumed to be wider than it is. For small velocities experimental observations confirm Newtonian mechanics to a high degree: it is only when velocities are large that discrepancies begin to emerge. Parity is nearly always conserved, but the class of phenomena for which it is conserved does not extend to cover everything.

Even if an experiment is well done, well designed and carried out within the range of applicability, the boundary conditions may be untypically wrong. We tacitly assume that the experiment will not be disrupted by a wandering comet or a black hole. We normally assume in chemistry that looking at the reaction will not interfere with it—although silver halides are known to be sensitive to light. Newtonian mechanics was developed without regard to electromagnetic phenomena, although now we know better than to ignore them. But still we have to ignore some conditions, or no experiment would ever be repeatable [30] We have to assume that exceptional circumstances do not obtain, in order to argue at all from the particular observations the experimentalist can make to the general propositions the theorist is interested in. But this assumption may always be wrong in the particular case, without any fault on the part of the experimentalist in his design or execution of the experiment, or in the theory that is being tested.

---

[30] See, more fully, *Space, Time and Causality*, ch.3, pp. 57ff.

For these reasons the opposition between empiricism and rationalism is not nearly so sharp as is often assumed. There is still
some tension, however. We have still to consider (1) what our
response should be if there is irresoluble conflict between theory
and observation—in the case of the Special Theory, if the speed
of light proved to be different in different frames of reference. We
have also to consider (2) what empirical content there is in our *a
priori* assumptions which enable us apparently to obtain scientific
results like rabbits out of a hat.[31] But these questions we still leave
unanswered for the present, and rest content with the fortunate
fact that so far as the speed of light is concerned our metaphysical
speculations are—in the end—borne out by physical experiment.

The argument of this section is further developed in §10.7

---

[31] See below, §4.5; and R.B.Angel, *Relativity: The Theory and its Philosophy*,
Oxford, 1980, pp. 110-115.

Chapter 2
Relative to What?

§2.1 Sameness and Order

This section presupposes the argument of §1.5

Einstein's approach was quite different from that given in the previous chapter. He was concerned with sameness rather than integration. He asked "Under what conditions will an electromagnetic wave appear the same to different observers?", "Will different observers reckon that an event occurred at the same time?", and "What must be the case if Maxwell's equations are to have the same form in different frames of reference?". It seems a much more down-to-earth approach. To discuss whether phenomena are the same is much less mysterious than to integrate intangible entities like space and time. And since the whole of science is concerned with observing and explaining similarities, nobody can take exception to Einstein's approach without calling in question the whole of the scientific enterprise.

Sameness is peculiarly characteristic of space. We think of space as being homogeneous, isotropic, and, usually, metrically amorphous. Every point in space is like every other point, every direction in space is like every other direction, and there is no naturally given spatial metric. Time and causality are not so samey as space. 'Before' is not the same as 'after', and causes are not the same as effects. Nevertheless, in other respects time and causality do manifest important samenesses. Time is homogeneous—origin-indifferent—and appears to have no natural metric; and it is characteristic of physical causes that they should be repeatable, that is, that the same cause should be followed by the same effect. For these reasons, as well as its being Einstein's own approach, it is natural to use sameness as the key to the Special Theory.

But there are difficulties. Sameness is not as simple as it seems. It seems to be a two-term relation, but is in fact a three-term one. The question "Am I the same as you?" cannot be answered until we have explained in what respect the sameness is to be assessed.

I speak the same language as you, and may well be the same nationality and the same sex, but am probably not the same age,
or the same height, or the same weight. Once we have specified
the respect—language, nationality, sex, age, height, or weight—we
can characterize *being the same as* by its formal properties. Formally, it is an "equivalence" relation, *i.e.* a relation which is (i)
transitive, (ii) symmetric, and therefore (iii) reflexive: if I am the
same age as you and you are the same age as James, then I am
the same age as James; if I am the same age as you, then you
are the same age as I am; and I am, of course, the same age as
myself. Any relation which is transitive, symmetric and reflexive
is an equivalence relation, and characterizes some sort of sameness
or similarity, different for different relations. In social life we are
familiar with the equivalence relations *contemporary, colleague, co-
religionist*, and *fellow-citizen*, and the corresponding equivalence
classes *cohort, peer-group, communion* and *nation*. In physics there
are many concepts which express some idea of sameness: *similar,
simultaneous, equal, equivalent, equipollent, isomorphic, isotropic,
homogeneous, uniform, featureless, amorphous, invariant, covariant, conservation* and *symmetry*. These concepts are usually best
understood in terms of the appropriate equivalence relation, which
are less abstract and more readily applied to empirical phenomena.

Besides equivalence relations there are ordering relations, typically expressed by the comparative, —*er than*, which also are transitive, but asymmetric and irreflexive: thus if I am taller than you,
and you are taller than Jane, I am taller than Jane, but if I am
taller than you, you are not taller than me, and I am not taller
than myself. The equivalence and ordering relations are the two
most important families of relations. As with the equivalence relations, there are many different ordering relations: I can be taller
than you, while you are cleverer than me. In the Special Theory,
however, whereas there are many different equivalence relations we
need to take account of, there is only *one* important ordering relation. Equivalence relations, although important, are always with
respect to some particular **frame of reference**[1] or other. But
the one ordering relation is absolute and the same for all frames of
reference, and thus plays a more fundamental role in the Special
Theory. The one ordering relation is that of *causal influenceability*,

---

[1] See below, §2.5.

or conversely *being a potential cause of*: influenceability is taken in a very wide sense; one event is causally influenceable by another not just if it actually is, or reasonably might be, influenced by it, but if it conceivably could. A.A. Robb, who pioneered the ordinal approach[2] used the word 'after', but that is a confusing usage. As we ordinarily understand it, the ordering generated by *after* is **strict**[3] that is to say, for any two events, the law of trichotomy holds, and either they are simultaneous or else one of them is after the other. It is this last assumption which, according to the Special Theory, does not hold for causal influenceability. The Special Theory does not deny the strict temporal ordering of events close at hand: in any frame of reference the time co-ordinate at the origin is strictly ordered; for every material particle we can consider its "proper time", [4] which, again, is a strictly ordered continuum. It is only for distant events that the ascription of dates is not straightforward. For these the simple common-sense concept of simultaneity has been replaced not by one but by two concepts, which are easily confused, and easily lead to inconsistency. On the one hand there is an *equivalence* relation, simultaneity properly so called, which is transitive, symmetric and reflexive, but always *with respect to* some particular frame of reference, and on the other hand there is an *absence* of an ordering relation, where the absence is obviously symmetric and reflexive, but not transitive. There is always an implication from the former to the latter: if two events are simultaneous with each other, then neither is causally influenceable by the other. For *strict* orderings, but only for them, there are converse implications: if neither of two events is causally influenceable by the other, then they are simultaneous with each other; if two events are each simultaneous with each other then if a third event is neither a potential cause of, nor causally influenceable, by the one, it is also neither a potential cause of, nor causally influenceable by, the other. But for other, non-strict orderings—

---

[2] A.A. Robb, *A Theory of Time and Space*, Cambridge, 1914, 2nd ed. 1936; and *The Absolute Relations of Time and Space*, Cambridge, 1921.

[3] Logicians often term strict orderings as linear orderings. It is a perfectly acceptable usage, but as we use the word 'linear' of functions in a different sense, we avoid using the word also to characterize orderings.

[4] See above, §1.5

partial orderings, as they are called—these latter implications fail: two events may be neither a potential cause of, nor causally influenceable by, each other, and yet not simultaneous either. And so the two different concepts, of *being simultaneous with*, and of *being neither before nor after*, fall apart. A tendentious example from outside physics may help to make the point clear. The relation *of being socially inferior to* is clearly an ordering relation, and is often taken to yield a strict ordering, with each person in society being either socially superior to some one else, or socially inferior, or else his social equal. But social equality means much more than that. It suggests a fairly wide range of shared interests and experience. An undergraduate in a British university is likely to acknowledge the Queen as his social superior and to think himself superior to a bookie, but to have no feelings of either superiority or inferiority towards an American Congressman. Yet he would not regard the American Congressman as his social equal, in the way he would regard another undergraduate as being one.

Simultaneity is like social equality. It tells us more than a mere absence of an ordering relation, but needs to be specified—with respect to some particular frame of reference —in a way that is not given us by the nature of things, but depends on our own arbitrary choice. Causal influenceability, by contrast, is absolute, and reveals the fundamental structure of spacetime as that of *light cones* which can be understood in terms of the metric developed in §1.5, but can also be viewed in purely ordinal terms.

§2.2 Light Cones and Topology

This section develops the metric of §1.5

Light cones can be defined for each point in spacetime in terms of causal influenceability, but are most easily thought of as the possible light rays through that point.[5] It is customary to represent them by drawing "spacetime diagrams" with the time axis going upwards, one spatial axis going sideways, and the other two not drawn. The light cone of any point divides spacetime into distinct regions. The point itself, the vertex of the cone, as a geometer

---

[5] See more fully §3.1.

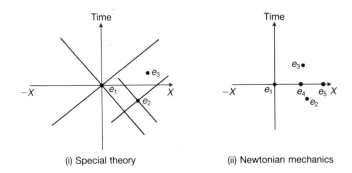

(i) Special theory   (ii) Newtonian mechanics

Figure 2.2.1 Time is pictured as going up the page. Of the three spatial dimensions, the X-axis is pictured as going from left to right in the plane of the page, the Y-axis is thought of as coming out of the page towards the reader, and the Z-axis is suppressed. The two events e_2 and e_3 are both of them neither before nor after e_1 according to this order, but cannot, both of them, be said to be simultaneous with e_1, since then they would be simultaneous to each other, and e_3 is clearly after e_2. In Newtonian mechanics, by contrast, there is no such possibility. Since there is no upper limit to the speed of propagation of causal influence, there are no events that in principle could neither influence nor be influenced by e_1, save those which happen at the same time and are thus "absolutely" simultaneous with e_1. If e_4 and e_5 are both of them neither before nor after e_1, then each is absolutely simultaneous with e_1, and so with each other.

would view it, constitutes the Here and Now for an observer at that point.

Inside the cone we have the Absolute Future and the Absolute Past: outside we have the rest of spacetime, the Absolute Elsewhere and Anywhen. The hyper-surface[6] of the light cone is the boundary between these regions: since no causal influence can be propagated faster than light, it delimits the ranges of possible causal influence. Any point in the Absolute Future can possibly be influenced by an event at the Here and Now: any point in the Absolute Past can

---

[6] Since the cone is really in three spatial, as well as one temporal, dimensions, we should refer to its surface as a *hyper*-surface, though often no error will ensue from omitting the 'hyper'.

possibly have influenced an event at the Here and Now: any point of the Absolute Elsewhere and Anywhen can neither be influenced by, nor have influenced, an event at the Here and Now.

ABSOLUTE
FUTURE

ABSOLUTE
PAST

Figure 2.2.2 Separations between HERE-and-NOW (marked O) and ABSOLUTE FUTURE (marked F) or ABSOLUTE PAST (marked P) *in* the interior of the light cone are timelike: $P \prec O$ and $O \prec F$.

The light cone indicates the different sorts of separation between points of spacetime and the Here and Now.[7] The separation between the Here and Now and points *in* either the Absolute Future or the Absolute Past (which we can symbolize by $P \prec O$ and $O \prec F$) is timelike; the separation between the Here and Now and points of the Absolute Elsewhere and Anywhen is spacelike (which we can symbolize by $O \parallel E$) and the separation between the Here and Now and points *on* the surface of the light cone is lightlike (Figures 2.2.2-4). [8]

The light cone thus indicates the fundamental metrical, and the rather peculiar topological, properties of spacetime. Topology is concerned with "nearness", points and sets of points that are close

---

[7] See above, §1.5.

[8] We use the prepositions 'within' or 'inside' for events in the interior of the light cone, 'of' or 'outside' for those of the exterior, 'on' for those on the boundary, and 'in' for those that are either within the interior or on the boundary.

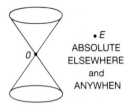

Figure 2.2.3 Separations between HERE-and-NOW and ABSO-LUTE ELSEWHERE and ANYWHEN (*i.e.* outside the light cone) are spacelike: $O \parallel E$.

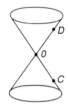

Figure 2.2.4 Separations between HERE-and-NOW and events on the hyper-*surface* of the light cone are lightlike: $C \longmapsto O$ and $O \longmapsto D$ **but** not necessarily $C \longmapsto D$.

together, that is those where the distance between them tends towards zero. In an ordinary space the distance between two points can be zero only if the two points are coincident, but in Minkowski space two points on the path of a light ray are not, according to our criterion, separated, even though they are, according to intuitive reckoning, a great distance apart. Hence whereas in an ordinary space two points are near only if the distance between them is tending towards zero, which can happen only when they are themselves actually coincident, in Minkowski space two points can be counted as being topologically near to each other without approximating in the least to being coincident. Minkowski spacetime thus has a topology quite different from that of an ordinary Euclidean space, and the light cone of each point has the peculiar property of being both very close to the point in question and constituting a physical barrier, an upper bound of possible causal influence. Topologically, Minkowski spacetime is very different from ordinary

Euclidean spaces, both in the small and in the large: and this difference is not only characteristic of it, but provides an approach to the
Special Theory which depends only on the light cones, themselves
definable in terms of causal influenceability alone.

Such an approach is very pure. It is also unfamiliar, and forbiddingly abstract. We therefore abandon purity for accessibility, and
consider first the easier and more traditional approaches depending on equivalence relations relative to various frames of reference,
rather than that depending on a unique ordering relation which
is in that sense absolute. Readers who spurn soft options should
turn to Chapter 3, and, having established the Special Theory absolutely, return to relative considerations in §2.3

The argument of this section is continued in §3.1

§2.3 Samenesses

There are many samenesses in physics. Space, in particular, and to
a lesser extent time and causality, are, we have seen, characterized
by a high degree of homogeneity and isotropy. Physicists often
make implicit appeal to assumptions about the samey-ness of space,
and, though caution is needed,[9] they are right to do so.

Einstein's insight, which led him to the the Special Theory, is
that certain things which are the same in one respect are the same
in another. There are essentially two samenesses, two respects that
have to be specified. On the one hand there are certain "frames of
reference", as they are called, [10] which are all equivalent in as much
as they are moving with uniform velocities with respect to one another, or "inertial" as it is termed: on the other there are certain
laws of nature, described with respect to, or having distinguishable
effects within, those frames of reference. Einstein, in his Principle
of Relativity, maintains that, in all frames of reference which are
the same in that they are inertial, the laws of nature are expressed
in the same form. This was not quite right. It is possible to fudge:
any law of nature can be given a covariant formulation,[11] and we

---

[9] See below, especially §8.4.

[10] See below, §2.5.

[11] E.Kretschmann, "Über den physikalischen Sinn der Relativitätspostulate,
A.Einsteins neue und seine ursprüngliche Relativitätstheorie", *Annalen der
Physik*, **53**, pp.575-614; see further below, §8.3.

need to be more careful in specifying what the sameness is that laws
of nature manifest in different frames of reference. Unfortunately,
it is difficult to give an entirely water-tight formulation of the sameness of laws of nature in different frames of reference, on account
of the many different sorts of co-ordinate system used.[12] Fortu-
nately, however, these complications are worrying only when we
are viewing the Special Theory from the standpoint of the General
Theory, and need to take curvilinear co-ordinates and accelerating
frames of reference seriously. So long as we are using only Cartesian co-ordinates and considering only inertial frames of reference,
we can accept Einstein's formulation as giving the essential thrust
of the Special Theory, though we shall later need to restrict it to
the laws of Newtonian mechanics and electromagnetism, and not
extend it, as Einstein did, to all laws of nature.[13]

Einstein's Principle of Relativity has many names. It is often
called the Galilean Principle of Relativity, or the Galilean Relativity Principle, because it was first formulated by Galileo with
respect to the laws of nature known in his day. What Einstein did
was to extend the second sameness, and maintain that not only the
laws of Newtonian mechanics but those of electromagnetism as well
were the same in all frames of reference which are equivalent in as
much as they are moving uniformly with respect to one another.
In order to stress the importance of this particular equivalence relation, we shall, except when we need to follow the usage of some
other author being discussed, speak of the Equivalence Principle.
But we need always to keep in mind the *range* of things said to be
equivalent (usually inertial frame of reference) and the *respects* in
which they are the same (usually the laws of Newtonian mechanics
and electromagnetism but sometimes, especially in Einstein's own
opinion, more).[14]

Different sorts of sameness are expressed by different equivalence
relations. Mathematically we can handle them in two ways: using

---

[12] For a full consideration and careful formulation see Michael Friedman, *Foundations of Space-Time Theories*, Princeton, 1983, ch.2, §2 and ch.4, §3; or
J.Earman, "Covariance, Invariance and the Equivalence of Frames", *Foundations of Physics*, **4**,1974, pp.267-289.

[13] See below, §3.9, p.118.

[14] See further below, §5.1, p.154,n.3, and §3.9, pp.117-119.

the terminology of the Theory of Groups which we are about to
introduce, we can say that certain transformations—change of origin, change of axes, change of unit—leave everything essentially
unaltered; or using the language of functions, we can say that certain physically important functions must not depend on certain
parameters representing position, orientation, or absolute units. A
set of transformations belong to a "group" if and only if [15]

(i) any two combine (in a way characteristic of the group) to
form a third which is also a member of the group,

(ii) their combination is associative,

(iii) to each transformation there is an inverse, and hence

(iv) the identity transformation is a member of the group.

A group of transformations can thus be seen as giving the "fine-
structure" of an equivalence relation.

Einstein's approach to the Special Theory was a group-theoret-
ic one. The fine-structure of the relevant equivalence relation
was given by as particular group of transformations—the Lorentz
group—and the sameness of laws of nature is expressed by their
remaining covariant under it.

§2.4 Types of Transformation

Physicists regard transformations in two ways: as active and as
passive. An active transformation expresses a relation between two
different phenomena, which resemble each other in the relevant respect; a passive transformation concerns only one phenomenon,
which is being described in two different ways. If I do two experiments, one in Oxford and one in Cambridge, but with the experimental set-up similar in all relevant respects, they are related
by a particular type of transformation called a "translation" or
"displacement": to each part of the Oxford apparatus corresponds
a piece of the Cambridge apparatus, situated 80 miles East North
East. Similarly if I repeat Tuesday's experiment on Thursday, there
is a temporal translation linking the two, in that to each process

---

[15] In recent years mathematicians and logicians have adopted the abbreviation 'iff' as short for 'if and only if': we shall make use of this convention in
formulae, though not in the text.

and event of the experiment on Tuesday there corresponds a similar one forty-eight hours later. These are "active" transformations. They are to be contrasted with passive transformations in which I change my way of describing, and in particular my way of *referring* to, the phenomenon; thus, if I had given all my spatial references to the Oxford experiment by means of the National Grid reference, and then decided to adopt instead an idiosyncratic scheme of my own, in which the base point (the origin, so to speak) was not off the coast of Land's End, but somewhere in North Cornwall, eighty miles East North East of the base point, then I should give different figures for locating the different parts of my experimental apparatus, but in spite of the difference of nomenclature the same experiment would be being described, with the same relevant correlations. I can say the same thing in English and in French and in German and in Latin. The words are different, but they are all true (or all false) together, and in spite of the manifest verbal differences, we reckon that what is really being said is the same in all cases. It is clear that the two aspects, active and passive, are linked. As the example suggests, the same transformation—a translation of 80 miles East North East—could equally well be a description, using the *same* scheme of reference, of a *different* experiment, one carried out in Cambridge instead of in Oxford, or a *different* description, using a different scheme of reference, of the *same* experiment, the one and the same Oxford experiment, but described in my idiosyncratic, instead of the normal National Grid reference, scheme.

In the Special Theory we are concerned with both active and passive transformations, because we are committed to two doctrines. We hold, first, that experimental observations should be *repeatable*. This is partly a matter of belief, but partly also a matter of choice. We believe, for a number of different reasons,[16] that the universe is fundamentally orderly and uniform. But in choosing to study physics we concern ourselves only with those features of phenomena that are very general. Physics is, by stipulation, the most general of all the sciences, and is seeking those laws that hold everywhere and "everywhen". Other sciences, biology or geology for example, are much more particular: but it is part of the definition of a physicist that he is concerned with the highest degree of gener-

---

[16] See below, §8.3, and §10.5

ality available; hence the emphasis on repeatability, which carries with it the requirement that if an experiment yields a particular result, then the "same" experiment, that is to say any experiment which is a transformation of that experiment by a temporal or spatial translation, should yield the "same" result. Hence there is built into physics a requirement of things working out the same under some sorts of active transformations—at least temporal and spatial translations.

There is also a doctrine about the ways we could, or at least should, refer to instants of time and points of space. Instants and points can have no intrinsic properties, as far as the physicist is concerned; else they would affect the results of experiments, and we should not have experiments at different dates or in different places always of necessity yielding the same results. All instants and all points are alike in some fundamental respect, and therefore they cannot possess any differentiating properties of their own, and can be individuated only by reference to other instants and points. From this it follows that it cannot matter much what point we choose as origin for a co-ordinate system or in what direction we align the axes. Everything of importance to a physicist must be preserved under a passive transformation from any one such system to any other.

The two requirements, that physical laws should hold everywhere and everywhen, and that every description of spatial and temporal location is to some extent arbitrary, but the arbitrariness must not be allowed to signify, are clearly linked—they both express some intimation of the featurelessness, or homogeneity, of space and time—but need none the less to be kept distinct in our thinking. The former expresses a positive assertion about the uniformity of the universe: the latter a negative denial of about our powers of description.[17] It was the latter which led Einstein to put forward the claim that all electromagnetic phenomena should be described

---

[17] The distinction becomes important when we compare the Special Theory with the General Theory. In the General Theory we no longer assume that all points of spacetime are alike, or that an active transformation from one spatiotemporal location to another will leave all the physics the same. Einstein's General Principle of Covariance was a generalisation of the group of passive transformations which should leave physical laws couched in essentially the same form. See further below §§8.3,8.4.

in essentially the same way not only in all standard co-ordinate systems but in all inertial "frames of reference".

> The argument of this section is taken further in §4.2 and §5.2

§2.5   Co-ordinates and Frames

Many people are confused by the two terms 'frame of reference' and 'co-ordinate system' which seem to be used almost interchangeably, but with some un-obvious difference. The most obvious difference is that the former is a physicist's term, expressing some rigid physical object, like a laboratory or a rocket, while the latter is a term of mathematical art. This, indeed, is one difference, but one which, though intuitively acceptable, cannot be taken for granted in the Special Theory, where the concept of rigidity is itself under question.[18] Instead of identifying a frame of reference by means of a material object, we should view it more abstractly, as a particular *class* of co-ordinate systems, those that are equivalent to one another under the group of "rigid" motions, that is to say that differ from one another only in origin or direction of axes.

Formally, a frame of reference is a *class*—an equivalence class—of co-ordinate systems.[19] A co-ordinate system is a way of referring to places, times, things or events. It is a sort of language. In English we call 'Venice' what the Italians call '*Venezia*', the French '*Venise*' and the Germans '*Venedig*'. Only, whereas in natural languages place-names vary somewhat unsystematically, a co-ordinate system refers to positions, dates, or events, in a standard systematic way. Besides the National Grid Reference system, we can refer to places by giving their latitude and longitude. There are many different sorts of co-ordinate systems, but in the Special Theory, as in Newtonian mechanics, we use Cartesian co-ordinates. To establish a Cartesian co-ordinate system for a Euclidean three-dimensional space, $\Re^3$, we need to specify a particular point as origin and three mutually perpendicular directions to give three axes that are rectilinear—*i.e.* straight—and orthogonal— *i.e.* at

---

[18] See above, §1.1, p.4.

[19] See Graham Nerlich, "Special Relativity is not based on Causality", *British Journal for the Philosophy of Science*, **33**, 1982, p.366; or his "Simultaneity and Convention in Special Relativity", in Robert McLaughlin, ed., *What? Where? When? Why?*, Reidel, 1982,, §3, esp. pp. 141,142.

right angles to one another. These we call the X-, the Y-, and the Z-axes. We then can identify a point by giving three numbers, x, y, and z, which we may understand as being the distances, expressed in terms of some standard units, in the direction of each of the three axes, which it is necessary to travel from the origin to reach the specified point. In the same way, if we want to refer to events, as we do both in the Special Theory and in Newtonian mechanics, we can do so by taking time as a fourth co-ordinate, so that each event is specified by four numbers (x, y, z, t).[20] The date t is expressed in terms of some unit[21] and measured from some time origin t_0, the birth of Christ, the founding of Rome, the first Olympiad, or the Big Bang.

There are many Cartesian co-ordinate systems we can adopt, all equally good from our point of view, between which we have simply to make a choice. It makes no difference, when we are doing algebraic geometry, what origin or what axes we choose, because although the co-ordinates will be affected by our choice, the distances between points and the angles between lines will remain the same. Similarly in physics, it is open to us to choose the origin, all but one of the axes, and whether we adopt a right-handed or a left-handed system.[22] It is an *arbitrary* choice, which we can make however we like, because it does not really matter what choice we make. For the laws of nature, which are relations between events, are unaffected. Thus the statement of Newton's Second Law of Motion

$$\mathbf{F} = \frac{d\mathbf{p}}{dt} \tag{2.5.1}$$

is independent of our choice of origin, or direction of axes. Because

---

[20] For a full account of the subtleties in the use of time as a co-ordinate, see Peter Kroes, *Time: Its Structure and Role in Physical Theories*, Dordrecht, 1985, ch.II, esp. §2, pp.62-63; see also below, §2.12.

[21] Our methods of measuring spatial distances and temporal durations are not as simple as is commonly supposed, and depend on a number of assumptions seldom made explicit or justified, but they work well enough in practice and are not, at this stage, of crucial importance for the Special Theory. For an account of the presuppositions of measurement, see more fully, *Space, Time and Causality*, Oxford, 1985, ch. 6.

[22] On the question of right- or left-handedness. see more fully ch.7.

space and time are viewed as being completely featureless, they are seen as having no landmarks available for anchoring any scheme of reference. Any choice between them must be essentially arbitrary, and so every physically relevant feature and every physically relevant correlation must be independent of that choice. There must therefore be some requirement of *covariance* under *passive transformations*, as a consequence not of our view of the way things happen, but of the way we are able to describe them in approved physical terms. It is the same as our belief that we must be able to translate, that is passively transform, true statements in one ordinary language into correspondingly true statements in another. If it is true to say in French *Vienne est au nord-est de Venise*, it ought to be true to say in Italian *La Vienna é alla nord-oriente di Venezia*. Truth, at least some sorts of simple truth, is independent of language. And in the same way we hold that physical truth must be independent of different physical "languages" that differ only in origin or axes.

The Special Theory extends the range of schemes for referring to places, dates, things and events, that make no fundamental difference to the underlying physics. It is like maintaining that geographical truth can be expressed not only in all Romance languages (Latin, Italian, French, Spanish, Portuguese and Rumanian), but in any Indo-European language (German, Russian, Persian, Hindi *etc.*). We can say not only in French *Vienne est au nord-est de Venise* and in Italian *La Vienna é alla nord-oriente di Venezia*, but equally in German *Wien ist nord-östlich von Venedig.* The equivalence class of languages in which geographical truth can be expressed is wider than we first thought. It is the same with Newtonian mechanics. Newton's Laws of Motion take the same form, when expressed in terms of a second co-ordinate system moving at a uniform velocity with respect to the first. Not only does a change of origin or axes make no difference, but also a change from one co-ordinate system to another, whose origin is, when referred to in the first co-ordinate system, moving at a uniform velocity. It is in order to emphasize this point, that we have found it useful to have an intermediate concept, a "frame of reference" which is an equivalence class of co-ordinate systems that are *obviously* equivalent to one another in as much as they differ only in origin or direction of axes. The transition from one co-ordinate system to another belonging to the same frame of reference is de-emphasized, in order to concentrate attention on the change from one co-ordinate system to another that was *moving* with respect to it. It is not obvious that this change will make no difference to the expression of laws of nature: indeed, it is not in general true: it is true, however, both for Newtonian mechanics and for the Special Theory, that a change from one co-ordinate system to another moving with respect to it, but at a *uniform* velocity, will make no difference to the expression of those two theories: they can be formulated, so to speak, not only in any Romance language (those that differ only in origin or orientation of axes), but in any Indo-European language (which may also be moving at a uniform velocity with respect to some other Indo-European language). It is a surprising feature.[23] The term 'frame of reference' takes it for granted that changes of origin, orientation of axes, and unit are irrelevant, and thus underlines the fact that a change from one frame of reference to another also makes no difference even though they are moving at a uniform velocity with respect to each other, *i.e.* provided they are neither rotating nor accelerating with respect to each other. Such frames of reference are called "inertial" frames. They, too, constitute an equivalence class, but of a different sort—a class of frames of reference all moving at a uniform velocity with respect to one another. It is for this equivalence class that the laws of Newtonian mechanics and of the Special Theory are always expressed in the same form.

The equivalence class, though wider than was at first sight obvious, it is not indefinitely wide. It does not include transformations into accelerating or rotating frames of reference, nor transformations into curvilinear, non-orthogonal co-ordinates, nor those that do not preserve continuity. It is as though geographical truth were preserved in the same form under translation from one Indo-European language to another, but not under translation into, say, Ural-Altaic languages, such as Finnish, Turkish, or Hungarian, nor into Semitic languages such as Arabic or Hebrew.

Once the *caveats* about rigidity have been registered, it is convenient to revert to the physicists' practice of identifying frames of reference by means of a solid physical framework, and we shall standardly consider a laboratory at rest on earth and a rocket moving *away* in the X-direction (*i.e.* to the right) with uniform velocity

---

[23] See further below, §8.3.

v. Two further cautions are, however, needed. We need on occasion to distinguish a frame of reference (corresponding to, say, the Romance group of languages) from the particular co-ordinate system actually being used, (corresponding to French, or Italian, or Rumanian). Typically the physicist chooses a co-ordinate system whose origin and axes are such as to make the working come out as simply as possible. It is not always immediately obvious that such simplifications are legitimate, and the question is further discussed in Chapter 4.[24]

We also need to be wary of describing a frame of reference as an observer. Often, indeed, it is a helpful device to think of one observer on earth and another as an astronaut, the former having the earth as his frame of reference, and the latter the spaceship as his; and we shall make use of this graphic way of presenting a thought experiment. But it is easy then to be misled into thinking that the Special Theory supports phenomenalism, and is telling us that reality is somehow dependent on the observer, as Bishop Berkeley maintained. That is not at all the same as claiming, as the Special Theory does, that certain physical concepts and magnitudes are relative to a frame of reference. The "observer" in the Special Theory is an observation post, a particular point of view, not a person with a particular, necessarily private sense-experience.[25] It is also dangerously easy to think that an observer is somehow necessarily at rest. What is true is that a frame of reference is necessarily at rest with respect to itself. An observer most naturally chooses a frame of reference in which he is at rest, just as a Frenchman most naturally chooses to talk in French. But he does not have to. Just as a Frenchman may on occasion choose to speak in Italian, so an observer may on occasion choose to adopt a frame of reference that is moving with respect to himself.[26]

This section is presupposed by,
and further developed in,
§§4.2, 4.3., 5.2 and 5.2

---

[24] In §4.3. For a rigorous account that starts from the concept of a frame of reference, see Roberto Torretti, *Relativity and Geometry*, Oxford, 1983, ch.1, §1.4, pp.14-15.

[25] See below, §9.4.

[26] See further below, §4.3, p.137.

§2.6   The Galilean and Lorentz Transformations

The Galilean and Lorentz transformations are two rival rules for re-referring to events and re-describing natural laws in the two frames of reference moving at a uniform velocity with respect to each other. We can see them as two competing dictionaries for translating from French to German or *vice versa*. In one of them (the Lorentz dictionary) *Vienne* is translated *Wien* and *Venise* is translated *Venedig*, but in the other (the Galilean dictionary) *Vienne* is translated *Weimar* and *Venedig* is translated *Vendée*. It is clear that the latter dictionary may be less serviceable for some purposes, such as translating true statements about geography in one language into true statements in the other. Nevertheless, it may preserve some truths. *La langue parlée par les habitants de Vienne est l'allemand.* translates into a true statement whether we translate *Vienne* by *Wien* or *Weimar*, and *Es ist möglich, guten Rotwein in Venedig zu trinken* is true whether *Venedig* is rendered by *Venise* or by *Vendée*. Although the deviant dictionary may be less good at translating geographical truths, it may be all right for translating truths about gastronomy or native-language speaking: the dictionary determines what sort of truth can be translated, and if we know what sort of truth we want to translate, we can tell which dictionary to use. According to the Special Theory the Galilean transformation constitutes a less good dictionary than the Lorentz transformation: it is cruder, giving renderings that are not quite accurate, a little like Grimm's law, which is all right for translating gutterals in certain positions from Romance into Teutonic languages, but fails on other points, and is altogether inapplicable for translations into other branches of the Indo-European family. The Galilean transformation is adequate for ordinary purposes when we are transforming from one frame of reference into another moving uniformly at a fairly low velocity with respect to it, but is not good enough if we want absolute precision, and is altogether inadequate when transforming from one frame of reference to another moving at a high velocity with respect to it.

The Galilean transformation from one frame of reference to another moving at a uniform velocity v in the X-direction can be

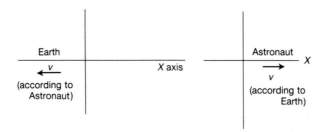

Figure 2.6.1 The Galilean transformation. If the astronaut's frame of reference—his spaceship—is moving away to the right from the Earth with uniform velocity v according to the frame of reference constituted by the Earth, whose X-axis points to the right, then according to the Astronaut the Earth is moving to the left with uniform velocity $v'(=-v)$, and the point $(x,0,0)$ at time t in the Earth's co-ordinate system will be described as the point $(x',0',0')$ at time t' in the Astronaut's co-ordinate system, where $x' = x - vt = x + v't$.

expressed algebraically as[27]

$$x' = x - vt$$
$$y' = y$$
$$z' = z \qquad\qquad (2.6.1)$$
$$t' = t.$$

In matrix notation this is

$$\begin{pmatrix} x' \\ y' \\ z' \\ t' \end{pmatrix} = \begin{pmatrix} 1 & 0 & 0 & -v \\ 0 & 1 & 0 & 0 \\ 0 & 0 & 1 & 0 \\ 0 & 0 & 0 & 1 \end{pmatrix} \begin{pmatrix} x \\ y \\ z \\ t \end{pmatrix} \qquad (2.6.2)$$

which may be written symbolically

$$X' = G(v)X \qquad\qquad (2.6.3)$$

---

[27] Normally, physicists use dashes or primes to distinguish (x',y',z',t') from (x,y,z,t), but these are easily missed. The non-numerate reader may find it helpful to use different-coloured biro in his working, or to read §4.2 first.

The Galilean transformation represents a very simple sort of dictionary. Almost all the names are unchanged: most importantly $t' = t$, which expresses the fundamental principle of Newtonian mechanics of the complete independence of time from space.[28] Except along the line of motion, other positions are unaltered: $y' = y$, and $z' = z$. The only change of place-name is in the direction of motion, where in order to obtain the co-ordinate in the Earth's frame of reference x, we subtract from the Astronaut's co-ordinate x' the further term vt, representing the distance travelled by the Astronaut's frame of reference with respect to that of the Earth. It is quite straightforward, but involves one consideration not yet noticed. Besides translating the place-names and dates, we need also to translate the velocity. For if the Astronaut's frame of reference is moving away to the **right** from the Earth's frame of reference with uniform velocity v, then the Earth's frame of reference is moving away to the **left** from the Astronaut's frame of reference with uniform velocity $-v$. If we keep to our convention and use primed letters for the Astronaut's language and unprimed letters for the Earth's language, then v', the velocity in the Astronaut's language used to describe the movement of the Earth's frame of reference, will be $-v$, the velocity in the Earth's language used to describe the movement of the Astronaut's frame of reference. Whenever we translate relative velocities from one frame of reference to the other, we have to change sign. It is very obvious, but can easily cause confusion. To avoid confusion, we shall, contrary to normal usage, distinguish v' from v, and use the former, rather than the latter with its sign reversed, in those formulae and matrices which are generally in primed terms. Essentially we are distinguishing between *un français-allemand dictionnaire* published in France and *ein französisch-deutsches Lexikon* published in Germany, and registering the fact that *l'autre langage* in a French dictionary refers to German, the language spoken east of the Rhine, and *die andere Sprache* in a German dictionary refers to French, the language spoken west of the Rhine.[29]

We have already come across a simplified form of the Lorentz

---

[28] See below, §8.7.

[29] See further below, §4.2, and table 4.2.1, where v is written V and v' is written **v**.

transformation;[30] written out in full, and again using the abbreviation

$$\gamma = \frac{1}{\sqrt{1 - v^2/c^2}}, \tag{2.6.4}$$

it becomes

$$\begin{aligned}
x' &= \gamma(x - vt) \\
y' &= y \\
z' &= z \\
t' &= \gamma(-x\frac{v}{c^2} + t).
\end{aligned} \tag{2.6.5}$$

If we express this in matrix form, the comparison with the Galilean transformation becomes clearer.

$$\begin{pmatrix} x' \\ y' \\ z' \\ t' \end{pmatrix} = \begin{pmatrix} \gamma & 0 & 0 & -\gamma v \\ 0 & 1 & 0 & 0 \\ 0 & 0 & 1 & 0 \\ -\gamma\frac{v}{c^2} & 0 & 0 & \gamma \end{pmatrix} \begin{pmatrix} x \\ y \\ z \\ t \end{pmatrix} \tag{2.6.6}$$

which may be written symbolically

$$X' = L(v)X \tag{2.6.7}$$

When $v = 0$, $\gamma = 1$, and both the Galilean and the Lorentz transformation become the Identity transformation

$$I = \begin{pmatrix} 1 & 0 & 0 & 0 \\ 0 & 1 & 0 & 0 \\ 0 & 0 & 1 & 0 \\ 0 & 0 & 0 & 1 \end{pmatrix} \tag{2.6.8}$$

It may easily be verified that $G(v)$ and $L(v)$ satisfy the symmetry conditions

$$G(v)G(-v) = 1 \quad \text{and} \quad L(v)L(-v) = 1 \tag{2.6.9}$$

In order to establish that the transformations do, indeed, form a group we need also to show that any two transformations together constitute a third, and that the associative law holds. For

---

[30] §1.4, pp.13-14.

the Galilean transformation it is easily shown. The composition of two transformations $G(v)$ and $G(-v)$ is itself a Galilean transformation. The associative law evidently must hold for Galilean transformations, since

$$G(v_1)G(v_2)G(v_3) = G(v_1 + v_2 + v_3), \qquad (2.6.10)$$

and addition is associative. For the Lorentz transformation the working is more cumbersome. The composition of two transformations is, as we should expect from Chapter 1,[31]

$$L(v_1)L(v_2) = L\left(\frac{v_1 + v_2}{1 + \frac{v_1 v_2}{c^2}}\right). \qquad (2.6.11)$$

This is enough to show that the combination of two Lorentz transformations is itself a Lorentz transformation. To verify the associative law for the Lorentz transformation we have to calculate

$$L(v_1)L(v_2)L(v_3) = L\left(\frac{v_1 + v_2 + v_3 + \frac{v_1 v_2 v_3}{c^2}}{1 + \frac{v_2 v_3}{c^2} + \frac{v_3 v_1}{c^2} + \frac{v_1 v_2}{c^2}}\right), \qquad (2.6.12)$$

which makes it clear that it does not matter in which order we associate $L(v_1)$, $L(v_2)$ and $L(v_3)$. Hence the associative law holds here too, and we have established that both the Galilean and the Lorentz transformations are groups.

The two transformations are very similar, but differ in two important respects:

(i) whereas in the bottom row of the matrix representing the Galilean transformation all the terms except the last are zero, thus making time altogether independent of position or velocity or anything else, in the bottom row of the Lorentz transformation there are other non-zero terms thus making t' depend not just on t, but also on x, the co-ordinate specifying the position;

(ii) the Lorentz transformation introduces a factor γ, depending on v, into the equations, whereas the Galilean transformation introduces no such factor, and is, in consequence, free of the puzzles associated with the Lorentz contraction and the dilation of time. The value of γ, $(1 - v^2/c^2)^{-\frac{1}{2}}$, depends not only on v, but

[31] §1.2, p.7.

on c. As c tends to infinity the Lorentz transformation tends towards the Galilean transformation, for then v/c tends to 0, and $1 - (v/c)^2$ tends to 1, which means that all the factors are unity, and the contribution of x to t' in the fourth equation of the Lorentz transformation is zero. The Galilean transformation can thus be seen as a *special case* of the Lorentz transformation. It is what the Lorentz transformation would be if the value of c, the speed of light, were infinite: it is also what the Lorentz transformation approximates to when v is small in comparison with c. Since terrestrial speeds are usually very small in comparison with the speed of light, the Galilean transformations are a good, indeed usually a very good, approximation to the Lorentz transformation for almost all practical purposes.[32]

The Galilean and the Lorentz transformations are both reasonable candidates for being the correct rule [cf.beginning of §2.4] for translating references to events and descriptions of natural laws from one frame of reference to another moving at a uniform velocity with respect to it. Although the Lorentz is more general, the Galilean is simpler and accords better with our present-day—pre-relativistic—intuitions. For physicists the most important difference between them is the theories they leave expressed in the same way. The Galilean transformation leaves Newtonian mechanics expressed by the same equations. Thus the statement of Newton's Second Law of Motion

$$\mathbf{F} = \frac{d\mathbf{p}}{dt} \qquad (2.6.13)$$

is not only, as we have seen, independent of our choice of origin, or direction of axes, but also independent of which inertial frame of reference we choose, *so long as we use the Galilean transformation* to translate from one to another. The same is true of Newton's other laws. They are covariant under the Galilean transformation. The Galilean transformation is adequate, as a translation rule between frames moving at uniform velocities with respect to one another, for all of classical—that is to say Newtonian—mechanics,but not for electromagnetism. Grimm's law is adequate, as between Romance and Teutonic languages, for Newtonian-mechanical truth, but not for electromagnetic truth: if we use it, we shall speak only *Bahnhofdeutsch*, which is all right for the tourist wanting to find

---

[32] But not quite all. See further below, §2.10.

the platform for the train to *Hannover*, but not for the scientist wanting to discuss the nature of light. For Maxwell's equations are **not** *covariant under the Galilean transformation*. This had led physicists to expect that one frame of reference—the aether's frame of reference, so to speak—could be picked out as being at rest. Einstein, however, thought that it was inconsistent that absolute motion could not be detected by a mechanical experiment and yet should, according to electromagnetic theory, be detectable by an electromagnetic experiment, and was thus led to the Lorentz transformation by asking himself what transformation would leave Maxwell's equations the same.

The rival merits of the Galilean and Lorentz transformations depend in part on our assessment of Newtonian mechanics and electromagnetism as physical theories, partly on the metaphysical assumptions they enshrine about space and time. The fact that the speed of light is finite and the same in all inertial frames of reference is powerful empirical evidence in favour of electromagnetism and the Lorentz transformation, while the evident distinctness of time and space has always been felt as a telling conceptual argument in favour of the Galilean transformation, although the concomitant commitment to the instantaneous, and so discontinuous, propagation of causal influence, and hence to "Action at a Distance" has been a weighty argument against. The issue is not one of "relativity". It is not that electromagnetism is, and Newtonian mechanics is not, "relativistic": Newtonian mechanics is just as "relativistic" under the Galilean transformation as the Special Theory is under the Lorentz transformation. What leads us to prefer the Lorentz transformation is a general belief that we can better explain Newtonian mechanics electromagnetically than electromagnetism mechanically, and secondly the particular considerations that, provided we use the Lorentz transformation,

(i) the speed of light is unchanged in all inertial frames,

(ii) the null result of the Michelson-Morley experiment is immediately explained, and

(iii) we see that it is not possible to detect absolute motion—motion, that is, with respect to the aether—either by mechanical or by electromagnetic means.

This section is presupposed by, and further
developed in ch.4, ch.5, and ch.8.

§2.7 The Derivation of the Lorentz Transformation

In Chapter 1 the Lorentz transformation emerged from the unification of space and time as a way of seeing velocities as resembling angles in spacetime. In the group-theoretical approach of this chapter it arises as the most natural transformation satisfying certain conditions of symmetry and simplicity that accommodates the speed of light being finite and the same in all inertial frames of reference. Although, as we shall see in Chapter 5, there are many other, neater, ways of deriving the Lorentz transformation, it is helpful to have at this stage the following, conceptually simple, though not particularly elegant, derivation for the simple case in which we have chosen co-ordinate systems in which the axes are aligned in the same direction, the X- and X'-axes are along the direction of relative motion, and the origins chosen so that $(0,0,0,0)$ and $(0',0',0',0')$ coincide.[33] It is based on one postulate and two symmetry requirements:

(i) the speed of light is the same in all inertial frames of reference;

(ii) the transformation depends only on the relative velocity between the two frames of reference, and is symmetric with respect to them;

(iii) the transformation is a linear function, involving only first powers of the co-ordinates x and t.

We write the Lorentz transformation in the form

$$L(v) = \begin{pmatrix} A & B \\ C & D \end{pmatrix} \qquad (2.7.1)$$

where A, B, C and D are functions of v to be determined. The co-ordinates in the two frames are therefore related by

$$\begin{pmatrix} x' \\ t' \end{pmatrix} = \begin{pmatrix} A & B \\ C & D \end{pmatrix} \begin{pmatrix} x \\ t \end{pmatrix} \qquad (2.7.2)$$

or, written out in full,

$$x' = Ax + Bt$$

$$t' = Cx + Dt$$

---

[33] A.S.Kompaneyets, *Theoretical Physics*, Moscow, 1961, p.194.

The coefficients B and C interrelate x and t, so they must change sign with v. Thus

$$L(-v) = \begin{pmatrix} A & -B \\ -C & D \end{pmatrix} \qquad (2.7.3)$$

Since, by condition 2,

$$L(v)L(-v) = 1 \qquad (2.7.4)$$

$$\begin{pmatrix} A^2 - BC & -AB + BD \\ CA - DC & -CB + D^2 \end{pmatrix} = \begin{pmatrix} 1 & 0 \\ 0 & 1 \end{pmatrix} \qquad (2.7.5)$$

so that

$$A = D \qquad (2.7.6)$$

and

$$A^2 - BC = 1 \qquad (2.7.7)$$

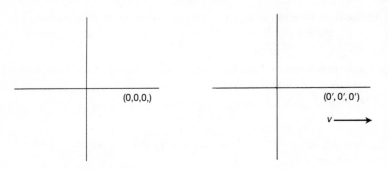

Earth $(0,0,0)$ Spaceship $(0', 0', 0')$

Figure 2.7.1 Two Frames of Reference

The spatial origin of the right-hand frame of reference, the Astronaut's spaceship, $(0', 0', 0')$, moves with velocity v to the right with respect to the left-hand frame of reference, the Earth, with spatial origin $(0, 0, 0)$.

The origin of one frame of reference moves with velocity v in the other frame of reference (see Figure 2.7.1), so that if $x' = 0$, $x = vt$. Thus

$$Avt + Bt = 0$$

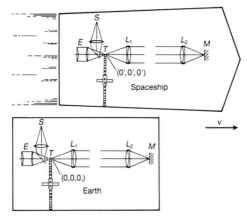

Figure 2.7.2 Measuring the speed of light. The speed of light is measured in the laboratory on Earth (left) by means of Fizeau's experiment (diagram taken from George S. Monk, *Light — Principles and Experiments*, McGraw Hill, 1937, p.115). If it were measured by a similar experiment carried out in a spaceship (right) moving with a uniform velocity with respect to the Earth, the result would be the same. So, although miles and seconds in the left-hand frame of reference are not the same as miles and seconds in the right-hand frame of reference, their ratio must be the same in order that c shall be the same in both frames of reference.

i.e.

$$B = -Av \qquad (2.7.8)$$

The invariance of the speed of light in the two systems requires (see Figure 2.7.2)

$$\frac{x}{t} = c = \frac{x'}{t'} \qquad (2.7.9)$$

Using (2.7.2) this becomes

$$c = \frac{Ax + Bt}{Cx + Dt}$$

which in turn yields

$$c = \frac{Ac + B}{Cc + D} \qquad (2.7.10)$$

whence, using (2.7.6) and (2.7.8)

$$Ac - Av = Cc^2 + Ac$$

and hence

$$C = -\frac{Av}{c^2} \qquad (2.7.11)$$

substituting back in (2.7.7) and using (2.7.8) gives

$$A^2 - \frac{A^2 v^2}{c^2} = 1 \qquad (2.7.12)$$

so that

$$A = D = \frac{1}{\sqrt{1 - v^2/c^2}} \qquad (2.7.13)$$

where the positive sign is taken to ensure that $x = x'$ and $t = t'$ if $v = 0$. It follows from (2.7.8) that

$$B = \frac{-v}{\sqrt{1 - v^2/c^2}}$$

and from (2.7.11) that

$$C = \frac{-v/c^2}{\sqrt{1 - v^2/c^2}}.$$

These expressions are clumsy, and it is convenient once again, as in the previous chapter and the previous section,[34] to write

$$\gamma = \frac{1}{\sqrt{1 - v^2/c^2}}$$

and

$$\beta = \frac{v}{c},$$

so that the matrix can be written succinctly

$$L(v) = \begin{pmatrix} \gamma & -\gamma v \\ -\gamma v/c^2 & \gamma \end{pmatrix} \qquad (2.7.14)$$

which is easier to read than when written out in full

$$L(v) = \begin{pmatrix} \dfrac{1}{\sqrt{1 - v^2/c^2}} & -\dfrac{v}{\sqrt{1 - v^2/c^2}} \\ -\dfrac{v/c^2}{\sqrt{1 - v^2/c^2}} & \dfrac{1}{\sqrt{1 - v^2/c^2}} \end{pmatrix} \qquad (2.7.15)$$

[34] See §1.3, p.12, equation (1.3.2) and §2.6, p.48, equation (2.6.4)

In fact we can achieve greater symmetry if we define the Lorentz transformation and the co-ordinates in terms not of t, but of ict: in that case the Lorentz transformation takes the equivalent but more symmetrical form is

$$\mathcal{L}(v) = \begin{pmatrix} \gamma & i\beta\gamma \\ -i\beta\gamma & \gamma \end{pmatrix} \tag{2.7.16},$$

so that

$$L(v)\begin{pmatrix} x \\ t \end{pmatrix} \equiv \mathcal{L}(v)\begin{pmatrix} x \\ ict \end{pmatrix} \tag{2.7.17}$$

If we write $\beta = \tanh\phi$, then $\gamma = \cosh\phi$ and, once again, we have the Lorentz transformation

$$\mathcal{L}(v) = \begin{pmatrix} \cosh\phi & i\sinh\phi \\ -i\sinh\phi & \cosh\phi \end{pmatrix} \tag{2.7.18}$$

as equivalent to a rotation in a complex plane. This is the form obtained in the previous chapter.[35] One of the properties of a rotation matrix is that it preserves the length of vectors on which it operates. Thus the spacetime interval

$$\sigma^2 = x^2 + y^2 + z^2 - c^2t^2 \tag{2.7.19}$$

is invariant.

The successive operation of two Lorentz transformations gives another Lorentz transformation, which yields the law of addition of rapidities. Thus

$$\mathcal{L}(\beta_1)\mathcal{L}(\beta_2) = \mathcal{L}(\beta_1 + \beta_2)$$

$$i.e. \begin{pmatrix} \gamma_1 & i\beta_1\gamma_1 \\ -i\beta_1\gamma_1 & \gamma_1 \end{pmatrix} \begin{pmatrix} \gamma_2 & i\beta_2\gamma_2 \\ -i\beta_2\gamma_2 & \gamma_2 \end{pmatrix}$$

$$= \begin{pmatrix} \gamma_1\gamma_2(1 + \beta_1\beta_2) & i\gamma_1\gamma_2(\beta_1 + \beta_2) \\ -i\gamma_1\gamma_2(\beta_1 + \beta_2) & \gamma_1\gamma_2(1 + \beta_1\beta_2) \end{pmatrix}$$

$$\equiv \begin{pmatrix} \Gamma & i\beta\Gamma \\ -i\beta\Gamma & \Gamma \end{pmatrix} \tag{2.7.20}$$

---

[35] §1.4, pp.14-15.

so that taking the ratios of the upper elements, the law of addition of velocities is

$$\beta = \frac{\beta_1 + \beta_2}{1 + \beta_1 \beta_2} \qquad (2.7.21)$$

This is the law of composition of velocities already obtained in §1.3 and in §2.6 (equation 2.6.8). It satisfies the required conditions that $\beta \approx \beta_1 + \beta_2$ when $\beta_1, \beta_2 \ll 1$, and $\beta \to 1$ as $\beta_1 \to 1$ or $\beta_2 \to 1$

There are many other derivations of the Lorentz transformation, some based on empirical generalisations, others fairly *a priori* in character. Their very variety is significant, and we shall return to them in Chapter 5. But first we need to consider other perspectives on the Special Theory, more closely concerned with the name relativity.

The argument of this section is continued in ch.5

§2.8 Spacetime and Simultaneity

The argument of this section presupposes §§2.1 and 2.2

The Lorentz transformation differs from the Galilean transformation in having time depend on distance and speed. Whereas in the Galilean transformation

$$t' = t. \qquad (2.8.1)$$

so that time is the same in all frames of reference, in the Lorentz transformation

$$t' = \gamma \left(t - \frac{vx}{c^2} \right). \qquad (2.8.2)$$

This means that the temporal ordering of the Special Theory is only partial, and therefore we cannot define simultaneity as simply being *neither before nor after*. But we *want* to define simultaneity none the less. For the temporal ordering of events that are close at hand is still strict; and we suppose that distant events which are themselves close to one another likewise have a strict temporal ordering; in which case we ought to be able to map their temporal ordering onto ours in an order-preserving way. So we *ascribe* dates to distant events, thereby deeming them simultaneous with other events, both distant and close at hand, to which we ascribe the same dates. But, in contrast to Newtonian mechanics, there

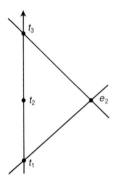

Figure 2.8.1 The Radar Rule. If we transmit a signal at t_1 and it is reflected (the event e_2) and is received at t_3, then we are to assign to the event e_2 the date t_2, where $t_2 = \frac{1}{2}(t_1 + t_3)$

are many different canons of simultaneity, and thus many different dating systems, which we can adopt.

We find it very difficult to accept that time, and so simultaneity, should not be something absolute, and many thinkers have rejected the Special Theory on the ground of its being contrary to common sense. Einstein criticized "common sense" on epistemological grounds. He asked how we could ascribe dates to distant events, given that there were no ways of transferring information faster than the speed of light. If we want to date distant events, all we can do is to send light-, or wireless-, signals to them, and receive back again the signals they send or return. Suppose a signal e_1 is emitted at date t_1, is received as event e_2, and immediately reflected back to the point of emission, where it is received at date t_3 (event e_3). What date should we ascribe to event e_2? Einstein argued that it would be inconsistent to ascribe to e_2 a date before t_1 or after t_3. For e_2, being an effect of e_1, must have a later date than the date of e_1, that is, t_1, and therefore of any t_0 that is before t_1; and being a cause of e_3, must have an earlier date than t_3 the date of e_3, and therefore than any t_4 that is after t_3. To avoid inconsistency we must ascribe to e_2 a date between t_1 and t_3: that is, if we ascribe to e_2 a date t_2,

$$t_2 = \varepsilon t_1 + (1 - \varepsilon)t_3 \qquad (2.8.3)$$

where $0 < \varepsilon < 1$. Einstein suggested that we should take $\varepsilon = \frac{1}{2}$ "by convention", and was thus led to propose a rule, which we can

describe anachronistically as the "Radar Rule", for ascribing dates
to distant events; the "Radar Rule" lays down that if we sent out
a radar pulse at time t_1 and it was reflected at a distant point and
was received again at t_3 (see Figure 2.8.1), we should ascribe to the
event, e_2, of the radar pulse's being reflected, the time t_2, where

$$t_2 = \frac{1}{2}(t_1 + t_3) \qquad (2.8.4)$$

and similarly we should ascribe its distance from the point of emission

$$r_2 = \frac{1}{2}c(t_3 - t_1) \qquad (2.8.5)$$

It is not self-evident that the Radar Rule gives the same results as the Lorentz transformation. In Chapter 4 we shall show
that in fact it does. Here our concern is, rather, with Einstein's
critique of simultaneity, and his suggestion that it was simply a
"convention" to set $\varepsilon = \frac{1}{2}$. It was a misapplication of the word
'convention'. There is, indeed, an element of conventionality is
ascribing *any* date to a distant event, granted the finite speed of
propagating causal influence, and the consequent partial ordering
of the causal relation: we are essentially *ascribing* dates to distant
events, not discovering them as something definite and given. But
this holds for any dating scheme, and not for the choice of $\varepsilon = \frac{1}{2}$
rather than, say, $\frac{1}{3}$ That choice is not properly characterized as a
convention. Proper conventions are ones where there is nothing to
choose between two alternatives—*e.g.* whether to use right-handed
or left-handed co-ordinate axes, or whether to regard multiplication as more or less binding, so far as omitted brackets go, than
addition—and we need to have a rule in order to be able to understand one another and concert our actions. Einstein was correct in
claiming that any choice of ε in the open interval (0,1) would be
consistent with the phenomena. But that is not to say that any
choice of ε would be as good as any other. The empiricist dogma
that empirical observations are the only grounds for choosing between theories, and that where two theories agree in the empirical
observations they predict there is nothing else to choose between
them, and any choice is arbitrary and a mere matter of convention,
is a dogma we should be rationally reluctant to accept without an
argument, an argument which in the nature of the case can hardly
be founded on empirical observation alone. Although there may

be no testable empirical consequences which can decide between one choice and another, there are considerations which favour the choice of $\varepsilon = \frac{1}{2}$ over all others. To say that $\varepsilon = \frac{1}{2}$ is to say that the speed of light is the same in both directions—that the radar pulse takes just as long to return from an event e_2 at a distance r_2 away from as it did to get there.[36] Of course we cannot check up on this experimentally. The whole philosophy of the Special Theory is that we cannot be in two places at once so as to see whether it really takes as long for the radar pulse to get back from e_2 as it did to get out there. We can only be where we are, and any ascription of dates or intervals to distant events or lengthy processes must be a construction from what we observe in our own patch. The one-way velocity of light is thus inherently likely always to elude empirical check. But we can think about it. And although we could construct a complicated theory in which there was a preferred direction, in which light went fastest, but all the other observational results came out the same as in Einstein's theory, it would be not only clumsy and counter-intuitive but would be against the main insight of the Special Theory, which is that there are no preferred frames of reference, and hence no preferred directions.[37]

The canons of simultaneity of different frames of reference that are not at rest with respect to each other will be different. It follows from the Lorentz transformation, but can also be seen to hold if we adopt Einstein's Radar Rule. Suppose an Astronaut

---

[36] It might seem also to be making the counter-intuitive assumption that the velocity of the source had no effect on the velocity of the radar pulse emitted from the source. This is not in fact an additional assumption. Once we have assumed that the speed of light is the same in all inertial frames of reference, we have already assumed that it is independent of its source.

[37] In view of further considerations to be adduced in the next chapter (§3.4, pp.103-105, only a very brief account has been given here of what has been a large controversy in the philosophy of science. For a full account see Wesley C. Salmon, "The One-Way Speed of Light", *Noûs*, **11**, 1977, pp.258-292. For an account of the Special Theory in which ε is not taken to be $\frac{1}{2}$, see John Winnie, "Special Relativity without One-Way Velocity Assumptions". *Philosophy of Science*, **37**, 1970, pp. 81-90, 223-238. The clearest statement of the original empiricist claim is given in Hans Reichenbach, *Space and Time*, New York, 1957, Ch.2, Sections 19, p.127, 22,p.144, and 27, p.168.

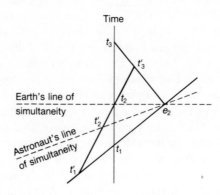

Figure 2.8.2 The lines of simultaneity on the spacetime diagram
connecting e_2 with t_2 on the Earth's time-axis and connecting e_2
with t'_2 on the Astronaut's time-axis, will be different.

is approaching the Earth with velocity v in the X-direction with
respect to the Earth, and suppose that before he reaches the earth
he emits (event e_1) at date t'_1 a radar pulse in the X-direction which
reaches a distant place (event e_2) and is immediately reflected back,
along the X-axis passing first the Astronaut and then arriving at
the Earth. The Astronaut and an observer on the Earth are each
assigning dates, t'_2 and t_2, to the same distant event, e_2, when the
radar pulse is reflected back towards them. On its return journey
it will reach the Astronaut at t'_3 and the Earth somewhat later at
t_3. By similar reasoning, the pulse on its outward journey will have
left the Earth at time t_1, having left the Astronaut at the *earlier*
date t'_1, since at that time the Astronaut had not passed the Earth,
and was further away from the event e_2 than was the Earth (see
Figure 2.8.3). According to the time-reckoning on Earth, the date
of the event e_2 is t_2, where $t_2 = \frac{1}{2}(t_1 + t_3)$: according to the time-
reckoning of the Astronaut the date of the event e_2 is t'_2, where
$t'_2 = \frac{1}{2}(t'_1 + t'_3)$. But t'_3 is before t_3 and t'_1 is before t_1, so t'_2 is
before t_2; hence the line t'_2—e_2 slants upward to the right, and is
rotated anticlockwise in comparison with the line t_2—e_2. Thus as
the world-line of the Astronaut, that is the line $x' = 0$, is rotated
clockwise, his lines of simultaneity, the lines parallel with $t'_2 = 0$

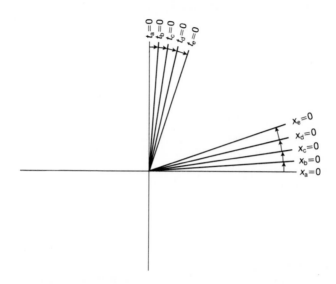

Figure 2.8.3 As the world-line is rotated clockwise, the hyperplanes of simultaneity, represented in the diagram as lines, are rotated anticlockwise. The explanation emerges in the next chapter (§3.3, Figure 3.3.2, p.96.)

are rotated anticlockwise (see Figure 2.8.3).

The argument thus given is geometrical, but can be established by non-geometrical argument. It seems rather counter-intuitive that the axes should rotate in opposite directions as the relative velocity of the two frames of reference increases; it can be explained as a consequence of our using hyperbolic functions in the Lorentz transformation. In §3.4 of the next chapter another, deeper explanation will emerge.

Einstein's critique of the common-sense concept of time and simultaneity does not really convince. It is like Leibniz' criticism of Newton's concept of Absolute Space and Absolute Time: although Newton cannot tell Leibniz how he can tell the absolute position of something, we are reluctant to conclude that there is therefore no difference at all. Verificationist arguments, if they worked at all, would cut too deep for intellectual comfort.[38] What Einstein

---

[38] See below, §9.3.

has really shown is that there is no single criterion of absolute simultaneity forced on us by the causal ordering of events. Instead, there are many canons of simultaneity that are compatible with the partial ordering given by causality, granted a finite maximum speed of propagation of causal influence. Which one we choose is not arbitrary, and is indeed "forced" upon us by our need to express the laws of physics in a coherent and covariant form. But it does depend on the frame of reference, and there are many possible inertial frames of reference that are, so far as electromagnetic theory is concerned, equivalent. Instead of absolute simultaneity, we have simultaneity with respect to a given frame of reference. Once we accept that *being at the same time as* is a triadic relation, and that we have to say not only what is the same as what but also in what respect, the fact that we have to specify in respect of what frame of reference we are relating events seems reasonable. Although it might have been the case that all inertial frames of reference yielded the same canon of simultaneity, as they do in Newtonian physics, it does not have to be the case; and if there is to be some integration of space and time into spacetime, with consequently some universal speed which is likely to be the upper limit for the propagation of causal influence, there will not be a universal strict ordering of dates given in the nature of the case, and there will be a variety of possible canons of simultaneity which we can without inconsistency adopt. It will not be an entirely arbitrary convention which ones we choose. There are good reasons for adopting the Radar Rule. But if we do, the canon of simultaneity will depend on the frame of reference chosen, and will be different for different ones moving with respect to one another with different velocities. Simultaneity, according to the Special Theory, is not given in the way that potential cause and effect is. It depends on our choice, the way we choose to *ascribe* dates to distant events—*i.e.* our frame of reference. But granted any one choice of a frame of reference, there is a natural canon of simultaneity it is rational to adopt.

§2.9   Simultaneity and the Present

The fact that simultaneity is a three-termed relation, and we always have to ask "simultaneous with respect to what frame of reference?" has given rise to some puzzles. It has been claimed that the Special Theory argues for a timeless view of time, because there is no absolute distinction between present, past and future.[39] If we consider two observers in two frames of reference moving with a uniform velocity with respect to each other, they will have different lines of simultaneity, and an event that is future to one will be past with respect to the other. In Figure 2.9.1 the dotted line ... represents the line of simultaneity of, say, an Astronaut, whose frame of reference is his spaceship, and the line of dashes – – – – – – represents the line of simultaneity of someone in a laboratory, whose frame of reference is the Earth.

The event E is both past and present, since it is simultaneous with an event F, say, which is itself simultaneous with the observer's 'Now'. So, it seems, we are forced to concede that there can be no inherent characteristic distinguishing the future from the present or the past, and that they are differentiated by nothing inherent but only their relation to particular observers. From this it is a natural step to adopt a "block view" of spacetime, and think of it as a timeless entity, in which all events, present past and future,

---

[39] H.Putnam, "Time and Physical Geometry", *Journal of Philosophy*, **64**, 1967, pp. 240-247; reprinted in H.Putnam, *Mathematics, Matter and Method. Philosophical Papers*, I, Cambridge, 1979, pp.198-205; C.W.Rietdijk, "A Rigorous Proof of Determinism Derived from the Special Theory of Relativity", *Philosophy of Science*, **33**, 1966, 341-344, and "Special Relativity and Determinism", *Philosophy of Science*, **43**, 1976, pp. 598-609; John W. Lango, "The logic of simultaneity", *Journal of Philosophy*, **66**, 1969, pp.340-350. For discussion of these claims, see Howard Stein, "On Einstein-Minkowski Space-time", *Journal of Philosophy*, **65**, 1968, pp.5-23; and "A note on time and Relativity Theory", *Journal of Philosophy*, **67**, 1970, pp.289-294; see also R.Sorabji, *Necessity, Cause and Blame*, London, 1980, pp.114-119; and R.Torretti, *Relativity and Geometry*, Oxford, 1983, pp.249-251. Part, however, of the argument put forward by Putnam and Rietdijk depends on the absolute nature of modal distinctions rather than a simple failure to appreciate the logic of simultaneity. See further below, §3.9.

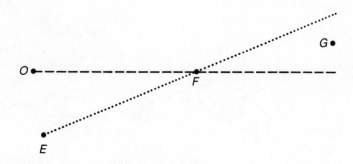

Figure 2.9.1   The event E is simultaneous with F which is itself simultaneous with O: so it should be simultaneous with O. But it is in O's past. So also the event G seems to be both future and past, future with respect to observers on Earth, and past with respect to the Astronaut.

already in some sense exist, and the apparent passage of time is nothing real, but only our becoming aware of what is already there, waiting for its moment of disclosure to us.

There is clearly something wrong with this account, plausible though it sometimes seems. It fails because simultaneity depends on the frame of reference. Simultaneity with respect to some particular frame of reference is an equivalence relation, so that if one event is simultaneous with another and that other is simultaneous with a third, then the first is simultaneous with the third. But simultaneity with respect to *different* frames of reference is not an equivalence relation at all. If I was at the same school as you, and you were at the same school as James, then whether it follows that I was at the same school as James depends on which school you shared with me and which with James. If you were at the same nursery school as I was, and were at Bristol Grammar School with James, nothing follows about my having been at the same school as James. Hence the argument fails: no event in the past of an observer can also be simultaneous with his Now; the argument purporting to show this depends on an equivocation in the use of the term 'simultaneous'.

But still, it may be objected, there is an inconsistent ascription of dates, without appeal to lines of simultaneity. In Figure 2.9.1 the event G is both future and past, future with respect to observers on Earth, past with respect to the Astronaut. The ascription of futurity, presentness, or pastness is, therefore purely relative, and indicates no real property of the event. But that argument also fails. The lines of simultaneity for a given frame of reference do not determine what is currently going on at distant places, but only what dates should be ascribed to them in order to make electromagnetic phenomena coherent. As far as electromagnetic phenomena are concerned, we have no means of telling exactly when a distant event—an event in the "Absolutely Elsewhere and Anywhen" as we have called it—takes place; but for any given frame of reference, IF we ascribe the same date to all events on a particular line of simultaneity, then the laws of electromagnetism apply neatly and yield harmonious results. So far as the Special Theory goes, simultaneity is a rather superficial and frame-dependent property, which we find useful for assigning dates to different events in different places, but which is not of fundamental importance in accounting for the propagation of causal influence. The ascription of presentness, pastness, or futurity, to events outside the light cone is nominal rather than real, and has no bearing on their ontological status.

This is not the end of the argument. If we ascribe a difference of modal status to what is present or past from what is future, it will be independent of our ascriptions of dates in some arbitrarily chosen frame of reference. Any God's-eye view of what is really fixed and unalterable as opposed to what is still open and indeterminate must impose a privileged frame of reference that is contrary to Einstein's principle that there are no privileged observers. In the next chapter we shall consider further the extent to which the Special Theory can accommodate superluminal velocities, and the bearing of the concepts of the Special Theory on the modal nature of time.[40] Meanwhile in this chapter we go on to consider further consequences of the Lorentz transformation which seem contrary to reason, and give rise to many perplexities, especially with regard to the dilation of time and the "Twins' Paradox". The Twins' Paradox seems to arise not so much from the dependence of simultaneity on the frame of reference (though, as we shall see, simultaneity is

---

[40] In §3.9.

involved too), as from the dilation of temporal intervals in frames of reference moving with respect to other frames.

§2.10 Time Dilation

According to the Special Theory, the apparent time-scale of events in a moving system is extended when compared with that of a stationary observer. More generally, observers in different inertial frames each reckon that the clocks of the others are running slow, by a factor of γ, *i.e.* $(1 - v^2/c^2)^{-\frac{1}{2}}$, where v is the relative velocity of the two frames. This is a logical consequence of the Lorentz transformation and like other consequences of the Special Theory it is also confirmed by quantitative experiments. If it were not so confirmed, we would have to think again, to see what had gone wrong with our arguments, or else abandon the Special Theory altogether.

Tests on the Special Theory give many experimental demonstrations of time dilation, and among these the observations of the decay of elementary particles are particularly striking. Many elementary particles, such as pions and muons, are unstable and after a very short time spontaneously decay into other, generally more stable, particles. The probability of decay per unit time of these particles is constant and so the number remaining from an initial population N_0 after a time t is given by

$$N(t) = N_0 exp \left(\frac{-t}{\tau} \right) \qquad (2.10.1)$$

where τ is the mean life. Thus if we measure the number of particles decaying in a given time from a large sample we can calculate the mean life to an accuracy that depends on the size of our sample. The decaying particles provide us with a natural clock.

If we now measure the mean life τ_0 of a population of particles at rest, and then τ_v, that of a population of similar particles moving towards us with velocity v, we can immediately check the correctness of the formula for the time dilation given by the Special Theory; if it is correct we should find that $\tau(v) = \gamma\tau_0$, so that the particles moving towards us appear to live longer.

One of the earliest demonstrations of this used the absorption of cosmic-ray muons as they come down the earth's atmosphere. The muon intensity is reduced by collisions in the atmosphere and by spontaneous decay. It was found that the reduction of intensity was greater in the atmosphere than in an equivalent mass of solid absorber, and this difference was attributed to the greater probability of spontaneous decay in the longer path through the atmosphere and this enabled the mean life of the muons to be found. Quantitative calculations are complicated by the energy and angular spread of the muons, but Rossi, using the known muon momentum spectrum, obtained good agreement with the time dilation as we should expect if the Special Theory is correct.[41]

Much more precise determinations are now possible using mono-energetic beams of particles from accelerators. In such an experiment, a beam of pions is passed through a series of electronic counters and the attenuation measured. The velocity of the pions is obtained from the time taken to go from one counter to the next. Combining these measurements and using the time dilation formula gives the mean life of pions at rest, which may be compared with direct measurements on pions brought to rest. A determination by Greenberg *et al.* carried out in this way, gave $\tau_v = 62.4$ ns for the moving pions and $\tau_0 = 26.03 \pm 0.04$ ns for the stopped pions. Using the value of $\gamma = 2.4$ for this experiment gave $\tau_0 = 26.02 \pm 0.04$ ns for the lifetime calculated from the moving pions. The closeness of these two values for τ_0 convincingly confirms the Special Theory.[42]

---

[41] B.B.Rossi, 1940, cited in next footnote.

[42] B.B.Rossi, "The Decay of Mesotrons 1939-1943", in L.M.Brown, ed. *The Birth of Particle Physics*, p.183. See also B.Rossi, N.Hilberg and J.B.Hoaf, *Phys.Rev.*, **57**, 1940, pp.461ff.; B.Rossi and K.Greisen, *Phys.Rev*, **59**, 1941, pp.223ff. A.J.Greenberg, *et al. Phys.Rev.*, **23**, 1969, pp.1267ff.

§2.11 The Twins' Paradox

The dilation of time seems deeply paradoxical, and has been rejected by many thinkers on that account alone. More sophisticated opponents have sought to derive consequences that are clearly inconsistent, so that the dilation of time can be rejected not merely for being surprising, but for being self-contradictory. The most famous of these apparent paradoxes is the "Twins' paradox", or in a more sophisticated version, the "Clock paradox".

There are two twins (or two clocks) initially at rest in the earth's frame of reference. One of them stays still, while the other is accelerated in a rocket to a speed very near that of light. According to the reckoning of the earth-bound twin the clock, and all the physiological processes of the twin speeding away in his rocket, are slowed down, so that time "goes more slowly" on the rocket. But conversely according to the reckoning of the twin on the rocket the clock, and all the physiological processes of the earth-bound twin, are slowed down, so that time "goes more slowly" on earth. No contradiction need arise, as we have seen, so long as they do not meet again. But what if they do? What if the speeding twin reverses the thrust of his rocket and slows down and then begins to move back to earth, again very nearly at the speed of light? During his return journey once again he will reckon that the clock, and all the physiological processes of the earth-bound twin, are slowed down, so that time "goes more slowly" on earth; the dilation of time depends on v^2, not v, so that the change in direction of relative velocity will not make any difference. Similarly the earth-bound twin will again reckon that the clock, and all the physiological processes of the twin speeding back home in his rocket, are slowed down, so that time "goes more slowly" on the rocket. In due course the one twin slows down and completes his journey and meets the other twin who had stayed on earth.

Each expects the other to be much younger than himself: they cannot both be right. We cannot fudge the issue by talking of distant dates being merely ascribed to events. The two twins are once again both together in the same place, at rest in the same frame of reference. They cannot each be younger, in the same place and in the same frame of reference, than the other: that is a plain contradiction, and if the Special Theory implies that, it is clearly inconsistent, and must be rejected.

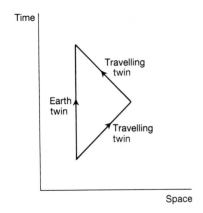

Figure 2.11.1 The Two Twins. The twin who goes out to the right, turns round, and comes back expects to find that his brother has aged much less than he has himself. His earth-bound brother expects to find him much less aged. Who is right?

It is easy to fault this purported *reductio ad absurdum*. There are many other assumptions the argument rests on, any one of which may be rejected rather than the Special Theory. But many of the assumptions we are inclined to reject turn out not to be essential for the development of the paradox, and by examining them in turn we are led to a more and more profound elucidation of the real nature of the paradox and a more and more careful exposition of its true solution.

The immediate response to the paradox is to protest that the two twins have different histories: one, the voyager in the rocket, has undergone either extreme or prolonged accelerations in changing from being at rest, in the earth's frame of reference, to going away from the earth at nearly the speed of light, and then in slowing down and imparting a great velocity in the other direction, and in finally coming to rest as he approaches the earth. Such violent changes are bound to affect his time-keeping processes, and so his physiological experience is not *pari passu* with that of the earth-bound twin whose time-keeping mechanisms have not been disrupted by acceleration and are therefore more to be believed. This is all right so far as it goes, but is open to two objections. In the first place, it suggests that it is the physical effect of the acceleration—the force used to accomplish it—that causes the trouble, in the same way as a shock can upset the correct running of a watch.[43] That difficulty can be avoided by arranging for the acceleration to take place in a very short time and switching off the clocks during that time. Secondly, and more seriously, no consistent temporal speeding-up effect can be ascribed to acceleration. Although in any one case it might be that the dilation due to the Lorentz transformation was off-set by an "anti-dilation" brought about by the acceleration, we have then to consider the parallel case in which the twin undergoes exactly the same acceleration but spends longer in unaccelerated motion. Clearly, the dilation effect is increased, so that even if the acceleration had exactly cancelled it out the first time, it would not do so the second, unless it depended not on the acceleration only but on other factors as well.

There may well be effects due to acceleration; but they are irrelevant. We can see this if we consider not the twins' paradox, but the clock paradox, which is a variant in which there are no accelerations at all, but simply three clocks passing one another at suitable times, and being synchronized as they pass. One clock stays at rest in the earth's frame of reference, and a second clock passes it, moving with respect to the earth's frame of reference at nearly the speed of light, and is synchronized as it passes, so that both these two clocks have their zero set at the time of passing. In the fullness of time the second clock passes a third clock, moving towards the earth, again with a speed nearly equal to that of light, and as they pass each other the third clock sets its clock to have the same reading at that instant as the second clock. In due course the third clock passes the earth, and compares its reading with that of the earth-bound clock. Again, a paradox seems to ensue. According to the earth-bound clock's reckoning, both the outward-bound clock and the returning clock have been running slow, and since the returning clock was set to have the same reading at the time of passing the outward-bound clock, the reading of the returning clock at the time of return, when it is in the same place as the earth-bound clock, should be earlier than that of the earth-bound clock itself. But this reckoning does not commend itself to either of the other two. Each reckons that the earth-bound clock has been

---

43 This suggestion was put forward in J.R.Lucas, *A Treatise on Time and Space*, London, 1973, §46, pp.234-235, and criticized by Graham Nerlich, in a review in *The British Journal for the Philosophy of Science*, **29**, 1978, p.298.

running slow throughout the whole journey, and so expects that it should tell an earlier time than theirs which were synchronized when they passed each other. Once again we seem to have a contradiction, which if not resolved would require the rejection of the Special Theory.

The clock paradox makes us examine more closely the transition from the frame of the outward-bound clock to that of the returning clock, and to construe this transition, not as a physical one involving dynamical effects such as forces, but as a conceptual one in which we shift from one frame of reference to another. It is in this shift that the apparent "missing time" of the earth-bound clock can be accommodated, so that although the earth-bound clock has indeed been running for a longer duration, much of it has escaped the notice of the reckoning of the other two clocks. We need to look more closely at the way the twins ascribe dates to each other.

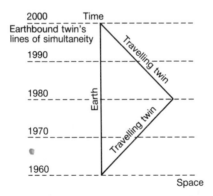

Figure 2.11.2 Spacetime diagram showing the earth-bound twin's lines of simultaneity

With the earth-bound twin there is no difficulty. He is in the same uniform motion all the time, and so his "lines of simultaneity" are the same throughout. If, as in Figure 2.11.1, his stationary progress is represented by a straight vertical line,[44] the lines of

---

[44] Is it fair to give the earth-bound twin the vertical world-line? Does not that beg the question in his favour? Why not draw another diagram with his world-line set at an angle to the vertical, and his lines of simultaneity correspondingly inclined (but at a contrary angle) to the horizontal?

simultaneity will be represented by parallel horizontal lines, indicated in dashes, thus: – – – – – –, as shown in Figure 2.11.2. It is easy to see how he ascribes dates according to his time-reckoning to events in the other twin's life. But what about the other twin's reckoning? He will have *two* sets of lines of simultaneity, corresponding to his two frames of reference.

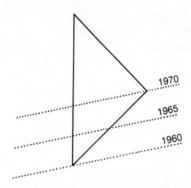

Figure 2.11.3 Travelling twin's lines of simultaneity on outward journey.

On the outward-bound journey (Figure 2.11.3), which we have represented by a straight line inclined clockwise to the vertical, his lines of simultaneity will be inclined anti-clockwise[45] to the horizontal, represented in the Figure by dots ....

On the return journey (Figure 2.11.4), which we have represented by a straight line inclined *anti*-clockwise to the vertical, his lines of simultaneity will be inclined *clockwise* to the horizontal, represented by dots and dashes · — · — · — · — If we put these two ascriptions of "travelling time" together, we see that there is a gap in the earth-bound twin's life that does not correspond to any duration of the travelling twin's life. Consider the date at which the travelling twin turns round: or in terms of the clock paradox, the date when the two moving clocks pass each other. What event in the earth-bound twin's life is thought by *the travelling twin* to be simultaneous with it? On the outward-bound journey it would seem to the travelling twin that a rather early event in his *younger* earth-bound brother's

---

[45] See above, §2.8, p.63, and Figure 2.8.3.

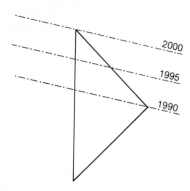

Figure 2.11.4 Travelling twin's lines of simultaneity on return journey.

life, say his marriage, would be simultaneous with the turn-round; according to those lines of simultaneity the earth-bound twin is, indeed, younger: after the turn-round, however, the calculation is quite different. The lines of simultaneity have altered in consequence of the alteration in the uniform state of motion of the travelling twin. The event in the earth-bound twin's life he would now take to be simultaneous with the turn-round is some event late in his life, say his retirement.

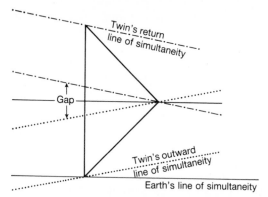

Figure 2.11.5 Travelling twin's lines of simultaneity on both outward and return journey.

Once again, the travelling twin will think that the earth-bound twin's clocks and physiological processes are slowed down, and will find support for that supposition in the fact that when they meet, the earth-bound twin will have aged relatively little *since retirement*, and will be, say, only seventy two, whereas the travelling twin reckons that he himself, has been quite a time on the way back. So, both on the way out and on the way back he thinks that his earth-bound brother is living life more slowly. But it does not follow that the earth-bound twin should be younger, because the travelling twin has left out of his calculations the years between the earth-bound twin's marriage and retirement which got overlooked in the shift of perspective from the outward to the return journey.

The essence of this solution is *kinematical* rather than *dynamical*. It is the conceptual shift in the lines of simultaneity, not any effect of the force required to bring about the needed accelerations and decelerations, that accounts for the mistaken reckoning of the travelling twin. He changes his time-reckoning in mid-course, not by moving round the hands of his clock, but by changing the rules for dating events on earth, and so naturally gets his calculations awry. The earth-bound twin has an uninterrupted view of what is happening to his travelling brother, and so his view of the matter is undistorted, while the view of the travelling twin is disrupted, and only seems to show, without actually doing so, that his brother must be younger than he is himself.

§2.12 Topology and Proper Time

The solution given in the last section is satisfactory so far as it goes, but does not assuage all our worries. It shows how the various ascriptions of dates to distant events, which we carry out in accordance with the Lorentz transformation so as to have the laws of electromagnetism the same in every inertial frame of reference, get out of step, but does not answer questions about *proper time*.[46] Proper time is not *ascribed*, but seems to be for real. How is it that the proper time the travelling twin has lived is less than that of his earth-bound brother?

The problem becomes more acute if we consider a case in which there is no alteration of motion, and so no shift of perspective due to change of inertial frame of reference. Suppose the universe is not infinite in all directions, but closed in one direction, and that

----

[46] See above, §1.5.

the travelling twin sets off in that direction, and in due course comes back to earth without having been accelerated, or in any way having changed his state of uniform motion in a straight line. In that case he will be entirely symmetrical with the earth-bound twin, and his view of the matter cannot be ruled out of court on the grounds of his having changed his perspective in mid-course. Each twin will have been in the same inertial frame throughout, and each will reckon that the other must be younger than he is himself, and a contradiction ensues.

It may be objected that if the universe is finite, so that a traveller setting off in one direction can come back on his starting place from the other direction without having ever departed from moving uniformly in a straight line, then space must be curved, so that any traveller will be subjected to gravitational forces, and will thus again not be *pari passu* with the earth-bound twin. But the curvature of space depends on the product of the linear curvatures in perpendicular directions. If the curvature in one of those directions is zero, then the curvature of the space as a whole is zero. Although a cylinder is in one sense obviously curved, in another sense its curvature is obviously zero: it can be unrolled flat, unlike the surface of a sphere. An undistorted map of a cylinder is perfectly feasible, whereas it is impossible to map a sphere faithfully onto a plane. Although a cylindrical spacetime would be unusual, and would run counter to the canon of isotropy, in as much as different directions were fundamentally different, it is possible, and would possess zero curvature so that the travelling twin could come back and compare notes with his earth-bound brother, without having been subject to any forces, gravitational or otherwise, which could disqualify his time-keeping from being reckoned quite as good and worthy of credence as his earth-bound brother's. And so the objection is ruled out and the apparent contradiction once again reinstated.

Although cylindrical spacetime is, indeed, flat, and so not open to objection on the score of being the scene of gravitational forces, it is not topologically the same as ordinary spacetime. It is not simply connected as ordinary spacetime is. In ordinary spacetime any closed curve can be continuously deformed to a point. In cylindrical spacetime, however, some curves, namely those that go right round the universe, cannot be shrunk to a point. They are in the same case as curves going right round a torus or anchor-ring. This constitutes the decisive difference between the two twins, enabling us to regard the time-keeping of one of them as canonical, and to discount that of the other as distorted. We can see this more clearly if we once again consider the lines of simultaneity of the travelling twin as he wends his way through the uttermost part of the universe. While he is still receding from the earth, his lines of simultaneity are parallel to those already drawn, and ascribe to youthful events in the earth-bound twin's life less youthful ones in his own: that is to say he thinks his earth-bound brother is younger than he is. But when he has passed the furthest point and is advancing towards earth from the other side, the same set of parallel lines of simultaneity will lead him to ascribe elderly episodes in his earth-bound brother's life as simultaneous with his own middle age. Although there has been no transition from one frame of reference to another by virtue of his having changed his state of uniform motion in a straight line, there *has* been a change in the direction he has to look towards in order to see the earth, and so a discontinuity in his ascription of dates to events on earth and thus also a gap in his account of his earth-bound brother's life. Where before he missed out the period of his brother's life between his wedding and his retirement on account of his changing his frame of reference, now he does so on account of his changing the direction in which he looks. He has, so to speak, crossed an "inter-galactic dateline" and on that account should recognise that he needs to adjust his dating system, just as the traveller on earth has when he crosses the 180^0 meridian. But this, though true is not completely self-explanatory. Why, we may ask, should not similar reasoning go through for the earth-bound twin? He also has to turn his head, and after having seen his brother disappear beyond α-Centauri, look towards Polaris to see him coming back from the other side of the universe. The turning of the earth-bound twin's head is as much a discontinuity as that of the travelling twin's head. Why should it not disqualify his time-reckoning as much as that of his brother? To answer this question we have to rely on a peculiar feature of the earth-bound twin's lines of simultaneity: they mesh; it does not matter which way he looks, he can see the same event at the opposite side of the universe. In Figure 2.11.2 this is represented by their being horizontal lines, which are perpendicular to, and therefore symmetrically disposed around, the vertical line that represents the earth-bound twin's world-line. In a spacetime with this peculiar topology we distinguish some frames of reference—

those whose lines of simultaneity mesh, as it were straight lines on the cylinder that are closed—which are genuinely at rest with respect to this spacetime from others which are not. Absolute rest has been re-introduced with respect to the *shape* of this non-isotropic space.

It is a peculiar feature of Newtonian space and the spacetime of the Special Theory that it should not have any distinguishing landmarks. We sometimes characterize Euclidean space and Lorentzian spacetime by saying that they are "flat", that is to say that they have zero curvature. This makes them particularly featureless, or metrically amorphous, so that all metrical features, and hence all physical quantities, are independent of spatiotemporal location, and space and spacetime are simply the arena in which things happen, and in no way an explanation of why they they happen. We are divorcing physics from geometry so far as possible, and trying *not* to have a science of geometrodynamics.

But the programme of non-geometrodynamics goes further still. We now see that we also need to specify that space and spacetime are simply connected, so that they shall be not only metrically amorphous, but topologically amorphous too. We want space and spacetime to be shapeless in a deeper sense than merely being flat, in order that it may be devoid of peculiar topological features, which might constitute landmarks for distinguishing some one frame of reference from all the others. Is this a reasonable want? Indeed, is the programme of non-geometrodynamics a sensible, or even a feasible, one? We need to be cautious. It is easy to let the programme of disconnecting space and spacetime from every sort of explanation run away with us.[47]   We cannot achieve absolute non-absoluteness because, ultimately, the concepts of sameness and difference are triadic, not dyadic, relations. The complete causal inefficacy of space and spacetime is therefore an unattainable ideal, and in the General Theory one that is altogether abandoned. In the Special Theory, however, as well as in Newtonian mechanics, we do seek to have spacetime, or space and time, as featureless as possible, and now we see that this is a topological, as well as a metrical, requirement.

Topology is conceptually important. It concerns the concept of nearness, on which our concepts of continuity and locality are

---

[47] See further below, §§8.3-8.5.

based. They are important both as a criterion of bodily identity and as a mark of causal connexion.[48] That spacetime should be simply connected is an assumption we readily make, but an important one, and one we need to make explicit none the less.

The connexion between topology and identity can be seen by considering an alternative response to the twins' paradox in cylindrical spacetime. The travelling twin need not acknowledge that he has returned to earth, but could maintain that he had reached a duplicate earth, a *doppleganger*, in which there was a qualitatively identical replica of the earth he had left behind. The universe on this construal would be like a vast palace of Versailles, with endless halls reduplicating one another indefinitely many times. We would have secured a simpler topology at the cost of ontological extravagance.

When we are dealing solely with material objects the decision between a simple topology with an extravagant ontology and a simple ontology with a complicated topology may be evenly balanced. But material objects are not the only entities we identify. Our criteria for personal identity are very different *pace* many modern materialists. Except for prisoners and hospital patients we are quite unable to tell whether the person we meet is bodily continuous with the person we met last time. Bodily similarity, rather than bodily continuity, is a prime method of identification, but not the only one. Often self-avowal is used, as on the telephone, and always we can have recourse to characteristics of personality and memories in addition to more overt, physical features. The travelling twin can therefore have further assurance that it is his brother he has come round to seeing again, not a replica in a far of region of space, because his brother remembers what they did together, and in particular the choices that he, the travelling twin, made in time past. The supposition that there were innumerable other replicas of himself who made the same decisions as he did at exactly comparable times commits him to a thorough-going determinism which undercuts all notions of personal identity and personal uniqueness. Further consideration of such a possibility belongs to a treatise on speculative metaphysics rather than one on the foundations of the Special Theory. If we have any reasonable concept of personal identity, we are constrained in the topology we can adopt, and hence

---

[48] See above, §1.1, and below, §10.5.

also in the topological shape we can reasonably ascribe to space, and thus, under the weird condition of cylindrical spacetime, the absolute equivalence between all frames of reference.

The twins' paradox argues against absolute relativity: it also argues for a spacetime approach, but one one that views time not primarily as a co-ordinate but a parameter.[49] We are led to this conclusion by a consideration of a further resolution of the paradox in terms of proper time. In ordinary, flat, simply connected, spacetime two straight lines cannot intersect in more than one point. If the twins are to meet again and compare notes, at least one of them must have followed a world-line that has a kink in it. A straight line was defined by Euclid as the shortest distance between two points. In Euclidean two-dimensional space we have the Pythagorean rule

$$s^2 = x^2 + y^2 \qquad (2.12.1)$$

or, in differential terms

$$\mathrm{d}s^2 = \mathrm{d}x^2 + \mathrm{d}y^2 \qquad (2.12.2)$$

together with the triangle rule that the distance between two points is less than or equal the sum of the distances between a third point and each of them. In the hyperbolic geometry of spacetime we have to replace the Pythagorean rule by the rule (taking $c = 1$)

$$\mathrm{d}\tau^2 = \mathrm{d}t^2 - \mathrm{d}x^2 - \mathrm{d}y^2 - \mathrm{d}z^2 \qquad (2.12.3)$$

and the triangle rule, as regards time-like separations, has to be expressed in terms of *less than* instead of *greater than* thus: the spacetime separation between two points is greater than or equal the sum of the spacetime separations between a third point and each of them. Just as in Euclidean space the effect of a kink in a path between two points is to make it longer than it would have been, so in the hyperbolic geometry of spacetime the effect of a kink in the world-line of the travelling twin is to reduce his space-time separation from his starting point. If we integrate along the

---

[49] For the distinction between parameter time and co-ordinate time, see Peter Kroes, *Time: Its Structure and Role in Physical Theories*, Dordrecht, 1985, ch.II, esp. pp.74-78.

paths of the two twins we obtain the total spacetime separation τ between their starting points—when they said "Goodbye"—and their final points—when they said "Good to see you again"—along their two world-lines, which represents the proper time that each has experienced. We are, essentially, comparing τ along the earth's world line with τ along the traveller's world line, (or, to be exact, the result of integrating $d\tau$ along the earth's world line with the result of integrating $d\tau$ along the traveller's world line,) and then we find that the former, because it is kinkless is longer than the latter. Just because the travelling twin has been going places, the total of the purely spatial component of his spacetime separation $dx^2 + dy^2 + dz^2$ is greater, which in view of the minus sign in the composite formula for spacetime separation $dt^2 - (dx^2 + dy^2 + dz^2)$ means that the total, which represents the proper time, is less. The Euclidean definition of a straight line as the shortest distance between two points is moderately unproblematic in a flat, simply connected space, but needs amendment in more complicated spaces. Instead of saying that it is absolutely the shortest (or longest), we have to gloss this with the word 'locally'. The travelling twin went the longest path he could, not in the sense that it would not have been longer if he had stayed at home, but that no slight variation would have made it longer: it was, granted the general route, maximal—an extremum—though not an absolute maximum in comparison to all the paths globally available.

The twins' paradox obliges us to take more account of proper time. What determines the ageing of each twin is the proper time τ that has elapsed along his trajectory, his world-line through spacetime. In Newtonian mechanics $\tau = t$ always: this is part of what Newton was trying to express when he said that absolute time flows evenly and equably from its own nature and independent of anything external.[50] We do not have to make it clear to ourselves whether we are considering time as the independent variable—parameter, as it is sometimes called—or as a dependent variable—co-ordinate. It all comes to the same in either case. In the Special Theory we are beginning to have to make the distinction. We assume, usually without thinking about it, that the trajectories of material bodies, moving with subluminal speeds, can be adequately represented by curves that always go forwards in time, and do not

---

[50] See below, §8.6.

ever loop. We do not need to distinguish the co-ordinate time constituted by the dates we ascribe to events and the parameter time constituted by the duration of proper time along a particular trajectory.[51] All we have to do in flat topologically respectable spacetime is to tilt the lines (*i.e.* hyper-planes) of simultaneity. In the General Theory when we are dealing with curved spacetime, we can no longer assume that the elapse of equal intervals of time along different trajectories will produce flat hyper-planes of simultaneity: and granted the weird topology of cylindrical spacetime, we find that if the twins traverse different trajectories, the concept of simultaneity is altogether discordant with that of the elapse of equal intervals of proper time. And it is intervals of proper time that are the same whatever inertial frame of reference we choose, and are the characteristic feature of subluminal, material bodies.

§2.13 Conclusion

We can now compare the approach of this chapter with that of Chapter 1. The theme of Chapter 1 is integration: the Lorentz transformation is an immediate consequence of the programme of integrating space and time; for if space and time can be integrated into a single spacetime, there must be some quasi-rotation between dimensions in spacetime analogous to ordinary rotations in ordinary space, with velocities resembling gradients and rapidities, or pseudo-angles, resembling angles. The Lorentz transformation emerged then as the natural analogue to ordinary rotations in Euclidean space.

The theme of this chapter, by contrast, is sameness. Sameness is always with respect to some feature or other, and each such feature determines a different equivalence relation and different equivalence classes. The Lorentz transformation emerges as the simplest transformation that would satisfy the "relativistic" requirements we are inclined to place on transformations so as to secure, what is already true of Newtonian mechanics, that the laws of physics shall be the same in frames of reference moving with uniform velocities with respect to one another. No further justification of this requirement

---

[51] It is this confusion that contributes powerfully to the appeal of the static picture of spacetime given by Weyl, quoted in §1.6, p.22 above.

is offered, or of why it should extend to frames of reference moving with uniform velocities with respect to one another and no further: the Equivalence Principle is justified on the basis of its working in Newtonian mechanics. The argument is simply "Same Again", taking account of the brute empirical fact, that the speed of light is finite. The Lorentz transformations follow, granted some very reasonable assumptions, and we are led then to acknowledge that sameness of time has to be specified with respect to some frame of reference, and the difference of perspective from one frame of reference to another leads us to foreshorten the spatial, and stretch out the temporal, components of spacetime separations.

The approach of the next chapter is based on a single ordering relation rather than on families of equivalence relations. It is an abstract approach, but captures some aspects of the Special Theory that other approaches miss.

Chapter 3
Absolute Approaches

§3.1 Causal Relations

This section picks up the argument of §§2.1,2.2.

Simultaneity is not a very fundamental property in electromagnetic theory. The previous chapter has shown that. What is to count as simultaneous depends on what frame of reference has been adopted, and two things can be simultaneous in one frame of reference and not in another, just as something can be at rest in one frame of reference and not in another.

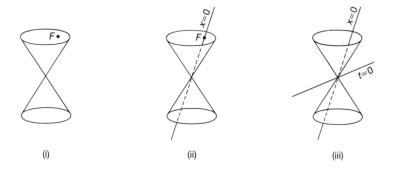

(i)   (ii)   (iii)

Figure 3.1.1 Given any future event F *inside* the light cone as in Figure (i), we can choose a frame of reference in which it is simply after the Here and Now, that is on the line $x = y = z = 0$ for that frame of reference, as in Figure (ii); corresponding to the given rest line $x = 0$ there is a hyper-plane (represented here as a line) of simultaneity, $t = 0$, as in Figure (iii).

It is desirable to characterize phenomena, so far as possible, in frame-independent ways, even at the cost of a high degree of abstraction. We therefore revert to the absolute approach outlined in §2.1 and §2.2 which is based on the ordering relation of *causal influenceability*, which, like the spacetime separation between two events, is the same in all frames of reference.

Light cones can be defined in three, closely related, ways. They can be characterized most intuitively, as their name suggests, in terms of light, or more generally electromagnetic radiation. Photons differ from other entities in having zero rest mass. The set of all possible paths of photons through each point constitutes a three-dimensional hyper-surface, which we call its light cone, and characterizes it in a physically significant way.

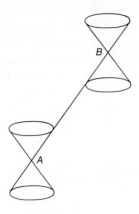

Figure 3.1.2 Light cones as sheathes: if a light ray can go from A to B, then B is *ON* the forward light cone of A, symbolically $A \longmapsto B$, and A is *ON* the backward light cone of B.

If A and B are such that a photon could go from A to B, we say that B is *ON* the forward light cone of A, and symbolize it by $A \longmapsto B$. Conversely we may say that A is *ON* the backward light cone of B.

More fundamentally, light cones can also be characterized in terms of causal influenceability. The relation "A can causally influence B", which we can symbolize by $A \prec B$, it picks out not just the hyper-surface constituted by the light rays, but the whole $(3 + 1)$-dimensional light cone; not the sheath but the sheaf. If A and B are such that A can causally influence B, we say that B is *IN* the forward light cone of A. Conversely we may say that A is *IN* the backward light cone of B.

Each event could have been influenced by any event either *IN* its Absolute Past or *ON* the earlier surface of the light cone, and may

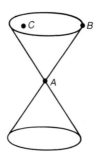

Figure 3.1.3 Light cones as sheaves: if A can causally influence B, and if A can causally influence C, then B and C are *IN* the forward light cone of A, symbolically $A \prec B$, and A is *IN* the backward light cones of B and C.

be able to influence any event either *IN* its Absolute Future or *ON* the later surface of the light cone, but cannot influence or be influenced by any event in its Absolute Elsewhere and Anywhen. The surface of the light cone is then defined as the boundary between those events that can be influenced or can influence and those that cannot.

The light cone is the boundary not only of the whole configuration but of its interior. The interior—the Absolute Future and the Absolute Past as we have defined it—is the set of those events that can be influenced or can influence by ordinary, non-electromagnetic means. It represents the world of Newtonian impacts. We can therefore base a third characterization of the light cone on the fact that the speed of light is **the upper limit of** causal propagation by means of entities with non-zero rest mass—by means of those objects that can abide being at rest. Such causal influence again is asymmetric and transitive, and therefore constitutes another ordering relation, here symbolized by $\ll$; it can clearly be defined in terms of the other two. What is not quite obvious, but nevertheless can be proved, is that each of the three relations can be defined in

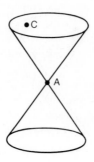

Figure 3.1.4 Light cones as interiors: if A can non-electromagnet-ically influence C, then C is *WITHIN* the forward light cone of A, symbolically $A \ll C$, and A is *WITHIN* the backward light cone of C.

terms of any one of the other two.[1] We are thus free to take either $\longmapsto$, *being on the path of a light ray from*, or $\prec$, *being causally in-fluenceable by* in the wider sense, or $\ll$, *being causally influenceable by* in the narrower sense, as our fundamental relation, and define the other two in terms of it. No particular one suggests itself on grounds of logic alone as more basic than the others—whichever one we start with, we need to define the others in order to develop a causal approach, because each has formal properties the others lack.

Which one we take as fundamental depends on our assessment of what is most fundamental from a physical or philosophical point of view. The physicist studying electromagnetism will take the light cone to be a sheath of light rays, all propagated at the same universal, invariant speed, reckoning this to be the fundamental feature of the Special Theory.[2] But $A \longmapsto B$ is not a transitive

---

[1] See E.C.Zeeman, "Causality Implies the Lorentz Group", *Journal of Math-ematical Physics*, **1**, 1964, pp.490-493; see also E.H. Kronheimer and R. Penrose, "On the Structure of Causal Spaces", *Cambridge Philosophical Society Proceedings*, **63**, 1967, pp.481-501; Roberto Torretti, *Relativity and Geometry*, Oxford, 1983, ch. IV, §4.6, pp.123.

[2] See below, §3.6 p.110 and §10.5, p.286.

relation: if I send a radar pulse to the moon and it bounces back again to earth, the separation between my sending it and receiving it back again is timelike, not lightlike. A philosopher inclined to materialism will start with $\ll$, *being causally influenceable by material bodies*, which he feels reasonably at home with, and will develop a relationist account of spacetime, much as earlier empiricists propounded a relationist account of space and time. But then in addition to the standard difficulties of the relationist approach, there is the awkwardness of a theory that shrinks back from the limit to which its concepts tend.[3] For these reasons we reject the first and third approaches, and take *being causally influenceable by* in the wider sense as the fundamental concept, though recognising that either of the others would be adequate from a logical point of view. Our aim is simply to develop a frame-independent topological account of the Special Theory, without in any way endorsing a reductionist ontology or epistemology.

§3.2   Zeeman: Causality Implies the Lorentz Group

This section develops §§1.5 and 2.2

The causal approach enables us to give a relational characterization of the light-cone structure of Minkowski spacetime, which in turn yields, surprisingly enough, its full metric. Before embarking on the main project, it is helpful to consider the more limited exercise of Professor E.C. Zeeman, who offers an interesting derivation of the Lorentz transformation from the relation of possible causal influence (in the narrower sense defined above, in which we consider only the interior of the light cone). He shows that provided Minkowski space has more than one spacelike dimension, the group of one-one mappings that preserve both the partial ordering relation and its converse (the "causal automorphisms") is the Lorentz group.[4] The key to his argument is the structure of light rays. They are shown to be straight lines and in a sense parallel to one another, and so give rise to the condition of linearity that we

---

[3] See above, §1.1, p.4.

[4] E.C. Zeeman, "Causality Implies the Lorentz Group" *Journal of Mathematical Physics*, **1**, 1964, pp.490-93.

needed for the first derivation of the Lorentz transformation in the previous chapter.[5]

Zeeman assumes that we have a Minkowski spacetime with one timelike dimension and more than one spacelike dimensions with Lorentz signature $(+ - \cdots -)$, and that we have a group of one-one mappings that preserve the relation *A can causally influence B by means of material bodies* symbolized by $\ll$. The proof is based on five lemmas. The first shows that what holds for causal influenceability in the interior of the light cone holds also for causal influenceability along light rays themselves. If f and f^{-1} are one-one mappings on Minkowski space, then they preserve the partial ordering $\ll$, if and only if they preserve the asymmetric non-transitive relation $\longmapsto$.

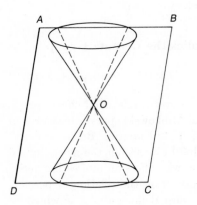 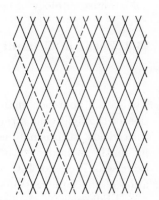

Figure 3.2.1 The intersection of the plane $ABCD$ containing the point O with the light cone of O is two straight lines as shown on the left: on the right we see the plane $ABCD$ as the two dashed lines in which the light cones whose vertices are in the plane generate two families of parallel lines.

From this there follows the second lemma that such a causal automorphism maps light rays to light rays, and hence, thirdly, parallel light rays to parallel light rays. In the normal case every two-dimensional subspace that is spanned by one timelike vector and

---

[5] Condition 3 on p.52 of §2.7

one spacelike vector will intersect the light cones in it in two families of parallel light rays, as in the diagram.[6]

The fourth lemma shows that because a causal automorphism
maps parallel light rays to parallel light rays, it must map them
linearly, provided there is more than one spacelike dimension. For if
there is more than one spacelike dimension,[7] there is more than one
pair of families of parallel lines available, and we can achieve any
translation along a light ray by compounding three suitable parallel
mappings. From this it follows that a causal automorphism maps
each light ray *linearly*. The fifth lemma shows that because again a
causal automorphism maps parallel light rays to parallel light rays,
it must map parallel equal intervals on light rays to parallel equal
intervals. It then follows fairly easily that it must be a Lorentz
transformation.

At first glance Zeeman does not seem to have achieved much. In
his presentation he defines $\prec$ in terms of a timelike separation, itself
defined in terms of a metric on Minkowski space; since he assumes a
Lorentz signature $(+ - \cdots -)$ in his definition of $\prec$, it seems hardly
surprising that he should then be able to derive a Lorentz transformation. But the definition of $\prec$ in terms of a metric relation
in Minkowski space is only one way of introducing the partial ordering relation of causal influenceability. We might start with that
as a primitive, and provided we can meet, or obviate, Zeeman's
requirement of adequate dimensionality, we should have derived
the Lorentz transformation from a partial ordering together with
a condition of "one-one-ness", but without having had to assume
linearity as in the derivation given in Chapter 2, or continuity. At
the very least Zeeman is able to make different assumptions from
those normally made in deriving the Lorentz transformation. The
continuity assumption, instead of having to be assumed directly, as
a consequence of linearity, is a consequence of the mapping's being
one-one and preserving the relation of causal influenceability, itself

---

[6] There is also a special case to be considered in which the two-dimensional
subspace is tangent to a light cone; the reader can find this discussed on
p.492 of Zeeman's paper.

[7] Compare the case of Desargues' theorem in projective geometry, which can
be proved only if there are more than two dimensions. See also below, §5.3,
p.164, §5.6 and §10.5, pp.285-286.

subject, as Hume saw,[8] to the requirement that it be continuous. The requirement of causality has been shifted from being an additional, and arguably arbitrary, assumption to being one defensible on general grounds. And although Zeeman himself introduces the relation of causal influenceability by reference to a standard metric defined on Minkowski spacetime, we can reasonably hope to characterize it simply as a particular sort of partial ordering. This was the approach of Robb.

§3.3 Robb's Axiomatic Approach

Early in the development of the Special Theory A.A. Robb presented a version based on one partial ordering relation.[9] Robb's work attracted little attention at the time, but has been more highly regarded in recent years, and offers some important insights into the nature of physical theory. It is very abstract. The reader unfamiliar with the Special Theory may well skip this and the next two sections at a first reading, and even the physicist may find Robb's axiomatic treatment too austerely mathematical for his taste.

Nevertheless there are virtues in Robb's approach. From the logician's point of view, ordering relations are the other great family of relations, which along with equivalence relations constitute the chief focus of interest in the logic of relations. Since the previous chapter was concerned with equivalence relations, considering which frames of reference were equivalent to one another, it is natural now to examine an approach that is based on an ordering relation. Moreover, whereas with equivalence relations we needed always to specify in what respect the different frames of reference were the same, Robb's approach makes use of only one ordering relation, that of causal influenceability, which has obvious and fundamental physical significance. Whereas it is a matter of choice what frame of reference we adopt, it is a matter of physical fact whether

---

[8] David Hume, *A Treatise on Human Nature*, Book I, Part III, Sections II and XV, pp. 75 and 173 in Selby-Bigge edition; see also J.R. Lucas, *Space, Time and Causality*, pp.40-42.

[9] A.A. Robb, *A Theory of Time and Space*, Cambridge, 1914; 2nd ed. (*Geometry of Time and Space*) Cambridge, 1936; and *The Absolute Relations of Time and Space*, Cambridge, 1921.

one event is causally influenceable by another or not. The light-cone structure generated by the relation of causal influenceability is part of the fundamental topology of Minkowski spacetime, not a particular standpoint we can take up if we please. From a purely logical point of view it would be preferable to develop this absolute ordinal approach before that based on a variety of equivalence relations, and, indeed, were it not for its much greater difficulty, we should have done that.

The absolute approach has further merits. Although it is primarily concerned with the topological structure of Minkowski spacetime, it turns out to have metrical implications. It thus shows how general qualitative assumptions can yield precise quantitative conclusions; and in particular, it enables us to give a definite answer to the problem, left unsolved in the last chapter, of the one-way velocity of light.[10] And the fact that the whole theory can be constructed on the basis of one ordering relation lends considerable support to relationist theories of spacetime. Nor should we be too much put off by Robb's austere axiomatic approach. After all, Newton argued *more geometrico*, like Euclid, in his *Principia*. The logical purity of the treatment, though far removed from the empirical approach of the practical physicist, illuminates the structure of the Special Theory from a new perspective.

Like Zeeman, Robb needs to make some assumptions about dimensionality, but he does not have to make any assumptions about the Lorentz signature of spacetime. The partial ordering relation itself is enough to distinguish timelike from spacelike directions, and though the axioms Robb needs to postulate are somewhat numerous, they are intuitively acceptable. His crucial step is to define what it is for a timelike and a spacelike direction to be "normal" or "orthogonal" to each other.[11] Orthogonality is the generalisation of the concept of perpendicularity. Two vectors are defined to be orthogonal to each other if their dot product is zero. In ordinary Euclidean geometry, two lines that are at right angles to each other

---

[10] See above, §2.8, pp.62-63.

[11] Robb uses the word 'normal'; see *Geometry of Time and Space*, Cambridge, 1936, pp. 209ff., or *The Absolute Relations of Time and Space*, Cambridge, 1921, pp.146ff.; but 'normal has many other uses, and so we use the word 'orthogonal' to avoid confusion.

are said to be orthogonal. In Minkowski spacetime it is possible to define orthogonality between a timelike line and a spacelike line in terms of the ordering relation, and hence orthogonality between two spacelike lines, from which congruence between line segments, and hence a metric, can in turn be defined.

Robb glosses the fundamental ordering relation as meaning 'after', and which we can interpret as 'causally influenceable by' (in the wider sense of §3.1) and regard as the converse of $\prec$. It ranges over possible events, $e_1, e_2, e_3, e_4, ... etc.$, and is dense, has no greatest or least term, and is a partial ordering throughout, that is, for every event e, there is another event e' such that it is neither the case that e is after e' nor that e' is after e. We can translate this condition into our terminology, using the converse relation $\prec$, $\neg((e' \prec e) \& (e \prec e'))$. Using this relation, Robb is able to pick out the paths of light rays, "optical lines" along the surface of each cone. These are characterized by their being maximal and hence unique and hence linear: e_1 and e_2 are on an optical line if, for all e_3 and e_4, such that $e_1 \prec e_3 \prec e_2$ and $e_1 \prec e_4 \prec e_2, either$ $e_3 \prec e_4$ *or* $e_4 \prec e_3$. These optical lines not only enable us to define, rather circuitously, planes and thus other lines, but provide us with families of parallel lines, in much the same way as in Zeeman's approach.

Robb uses an "Optical Parallelogram", or as Winnie calls it, a "Robb Rectangle", to define conditions under which a spacelike line and a timelike line are orthogonal. The definition illustrates the way spacetime diagrams succeed in capturing underlying truths. Two lines, one timelike (AD on Figure 3.3.1, the "The Optical Parallelogram") and the other spacelike (BC) are orthogonal if and only if they are the diagonals of a parallelogram ($ABCD$) whose sides are light rays. This explains why on the spacetime diagram (Figure 3.3.2) the axes rotate in opposite directions, as we noted in Chapter 2.[12]

In order to achieve these constructions, Robb, like Zeeman, needs more than one spatial dimension, which he secures with his Postulate XI. Thus in Figures 3.3.4 and 3.3.5, Postulate XI stipulates that if A and B are neither influenceable by the other, then if C is on the forward light cones of them both there is another possible

---

[12] §2.8, see Figure 2.8.3 on p.63.

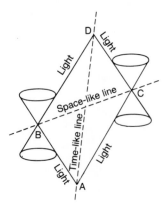

Figure 3.3.1 The Optical Parallelogram (or "Robb Rectangle"), in which light cones yield orthogonals. Winnie's name for this figure, the Robb Rectangle, is a good name, giving due recognition to Robb, but in order to constitute a rectangle, the paths of light rays must be at right angles in the spacetime diagram, which in turn requires that we take the value of c to be 1. In most cases this is unobjectionable, but since we want to emphasize how much Robb's treatment tells us about the geometry of spacetime diagrams without making any adventitious assumptions, we use light cones in which the light rays are not inclined at right angles to each other. Thus in spite of appearances, the lines AD and BC count as being orthogonal, since they join the vertices of intersecting light rays through B and C.

event, D, that is also on the forward light cones of them both, and similarly if E is on the backward light cones of them both there is another possible event, G, that is also on the backward light cones of them both.

Later postulates secure the existence of a third spatial dimension and that there are no more than three spatial dimensions, but this is not essential to his system of proofs generally.

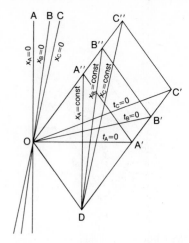

Figure 3.3.2 The optical parallelogram shows also why the lines of simultaneity and the rest-lines rotate in opposite directions. OA', OB' and OC' represent the lines $t_A = 0$, $t_B = 0$ and $t_C = 0$ of Figure 2.8.3. The optical parallelograms are $ODA'A''$, $ODB'B''$ and $ODC'C''$, whose diagonals are DA'', DB'' and DC'', representing rest-lines parallel to the world-lines $x_A = 0$, $x_B = 0$ and $x_C = 0$.

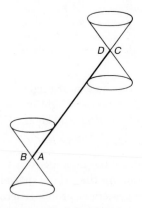

Figure 3.3.3 In the special case of the optical parallelogram (Figure 3.3.1) in which A coincides with B and C coincides with D, the light ray AC is the same as the diagonal AD, which is orthogonal to BC which is the same as AC. So AC is orthogonal to itself.

§3.4   Winnie's Presentation

Robb's work has been developed in recent years by Winnie, Tor-

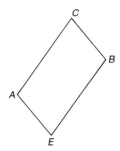

Figure 3.3.4 If there is only one spacelike dimension and $A \parallel B$, then there is only one possible event C such that $A \longmapsto C$ & $B \longmapsto C$; and similarly only one possible event E such that $E \longmapsto A$ & $E \longmapsto B$.

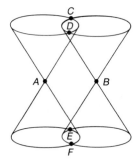

Figure 3.3.5 If there are more than one spacelike dimensions then the forward light cones of two events, A and B, neither of which is influenceable by the other, intersect in more than one point, such as C and D; similarly, both E and F are on the backward light cones of A and B. If, however, there is only one spacelike dimension, the light cones are simply pairs of straight lines, and intersect in only one future event and only one past event.

retti and Malament.[13] Winnie starts with a symmetric relation of

---

[13] J.A.Winnie, "The Causal Theory of Spacetime", in J.S. Earman *et al.*, eds., *Foundations of Space-Time Theories*, Minnesota Studies in the Philosophy of Science, Minneapolis, 1977, pp.134-205. Roberto Torretti, *Relativity and Geometry*, Oxford, 1983, ch. IV, §4.6, pp. 121-129. David Malament,

causal connectability, κ,[14] which we have defined in terms of causal influenceability by the condition:

$$a \; \kappa \; b \quad \text{iff} \quad a \prec b \vee b \prec a. \tag{3.4.1}$$

From this Winnie defines light-connectability. In general if two events are causally connected there will be others connected to them both and not causally connected with each other, as in Figure 3.4.1.

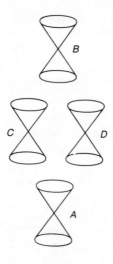

Figure 3.4.1 The general case in which C and D are causally connected to A and B, but not to each other.

But if the two events are on a light ray, then there is no latitude for two other events, if they are causally connected with them both,

"Causal Theories of Time and the Conventionality of Simultaneity", *Noüs*, XI, 1977, pp.293-300. An earlier treatment is to be found in H.Mehlberg, "Essai sur la théorie causale du temps. I and II. *Studia philosophica*, Lemberg, **1** and **2**,1935 and 1937, pp.119-260 and 111-231; tr. in H.Mehlberg, *Time, Causality and the Quantum Theory*, Reidel, Dordrecht, 1980.

[14] Winnie actually uses γ, but since we have used γ as an abbreviation of $\frac{1}{\sqrt{(1-v^2/c^2)}}$, we have replaced it by κ.

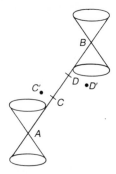

Figure 3.4.2 The special case, in which A and B are on a light ray. Then there are no points C' and D' connected to both A and B but not to each other. C' is causally connected to A but not to B, D' to B but not to A. Any points, such as C and D, that are connected to them both are on on the light ray from A to B, and therefore connected to each other.

not to be on a light ray, and hence causally connected to each other, as in Figure 3.4.2. So we can define light-connectability formally:

$$a \; \lambda \; b \; \text{ iff } \; (a = b) \vee$$
$$((a \; \kappa \; b) \; \& \; (\forall x)(\forall y)((x \; \kappa \; a \; \& \; x \; \kappa \; b \; \& \; y \; \kappa \; a \; \& \; y \; \kappa \; b) \to x \; \kappa \; y));$$
$$(3.4.2)$$

which lays it down that two events are light-connected if and only if either they are identical or they are causally connected and any two other events which are causally connected with them both must be causally connected with each other.

From light-connectability Winnie can define light rays, which are straight lines in Minkowski spacetime. Other straight lines are more difficult to characterize, but he uses an ingenious definition, due to Latzer,[15] which uses the triangle inequality to characterize spacelike straight lines. Three events are *collinear in a spacelike fashion* if and only if their light-cone hyper-surfaces fail to have a common intersection; that is to say there is no point on (as opposed to within) the light cones of all three. It is easy to see that if three points are not collinear then by the triangle inequality

---

[15] R.Latzer, "Nondirected Light signals and the Structure of Time", *Synthèse*, **24**, 1972, pp.236-280.

a light signal from one going directly towards the further of the others will overtake a signal going via the nearer one. Hence there must be some point that is light-connectable with all three: but if they are collinear, the light rays in any plane containing the three events will form two parallel families, and so there can be no point that is light-connectable with all three.

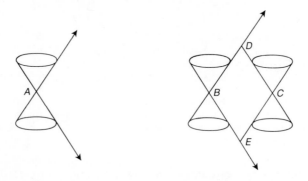

Figure 3.4.3   Light cones yield optical straight lines. If A, B, C are collinear, light from A will never overtake light from B in its path towards the world line of C. The only events common to the light cones of B and C are D in their Absolute Futures and E in their Absolute Pasts. But light from A cannot reach D, nor light from E reach A.

Collinearity for timelike lines is likewise defined by failure. Given any two events that are causally connected but not light-connected, there is a hyper-plane in which their light cones intersect. In general, given three events that are all separated from one another in a timelike fashion, the hyper-plane thus defined for any of them will intersect; but if, and only if, they are collinear the three hyper-planes fail to intersect.

Once linearity is defined, a spacelike straight line and a timelike straight line are said to be orthogonal to each other if and only if they are diagonals of an optical parallelogram.[16]

From this it is possible to define orthogonality between two spacelike straight lines. Two spacelike straight lines are orthogonal if and only if every timelike straight line that is orthogonal

---

[16] In Winnie's presentation, a Robb Rectangle.

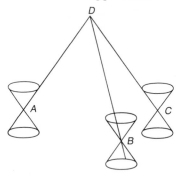

Figure 3.4.4 Following from Figure 3.4.3, if A, B, C are NOT collinear, there will be some point D, which is on a light ray through A, on a light ray through B and on a light ray through C.

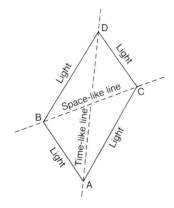

Figure 3.4.5 The optical parallelogram.

to one is also orthogonal to the other. Further work is still needed to define congruence on parallel, and then on non-parallel, lines, but it is plausible—and in fact true—to suppose that such a definition will be forthcoming. And so we find that starting simply from light-cone topology we have established a metrical structure on Minkowski spacetime without any further axioms or assumptions.

Modern readers find vectors easier to work with than pure geometry after the manner of Euclid. If we consider the ordered pairs of possible events A and B as vectors, we can use the definition of orthogonality to determine the conditions an inner product $< u, v >$ must satisfy. Since the inner product of two orthogonal vectors is zero, and since every lightlike line is orthogonal to itself, we have for, any vector u along the path of a light ray, $< u, u >$ $= 0$, and hence that the "length" of any vector u along the path of a light ray, namely $\sqrt{<u, u >}$, is zero; all lightlike vectors are null. From the Optical Parallelogram it follows that two vectors u and v are orthogonal if $u + v$ and $u - v$ are lightlike vectors, that is if $< u + v, u + v >= 0$ and $< u - v, u - v >= 0$. It follows that $< u, u >= - < v, v >$, which means that the ("length")^2 of time-like and of spacelike vectors is of opposite sign. It also follows that two timelike vectors cannot be orthogonal to each other, whereas two spacelike vectors can. We thus have an intrinsic characterization of lightlike, timelike and spacelike vectors, and the Lorentz signature. A lightlike vector is one that is orthogonal to itself: a timelike vector is one whose inner product with itself is opposite in sign to that of any vector orthogonal to it: and a spacelike vector is one that can be orthogonal to another vector that is orthogonal to one orthogonal to the spacelike vector itself. The Lorentz signature must be of the form $(+ - \ldots -)$ (or equivalently $(- + \ldots +)$). And then, using Zeeman's argument, we have the Lorentz transformation itself.

It is a remarkable result. We may still be sceptical about it—without giving the proofs in detail it is impossible to disarm scepticism fully—and wonder how it is that Minkowski spacetime can achieve what Euclidean space cannot. At first sight it would seem that the partial causal ordering of the absolute approach to the Special Theory was much messier and less powerful than the complete causal ordering of Newtonian mechanics. The reason why the opposite is the case is because the complete causal ordering has too little structure to enable us to characterize spacetime fully, so that we have to resort to arbitrary choices of what is to count as being at rest. In the Special Theory each rest-axis is correlated with a corresponding simultaneity axis, as in the three diagrams opposite (Figure 3.4.6), whereas in Newtonian mechanics all the simultaneity axes are the same, so that any one rest axis will correspond to the one and only simultaneity axis just like any other, as opposite (Figure 3.4.7).

Newton himself was aware that his mechanics gave no preferred frame of reference that was at absolute rest, and hoped that the centre of mass of the solar system would serve. What we now see is that the lack of structure in Newtonian space-and-time, $\Re^3 \times \Re^1$ is more profound. In $\Re^3 \times \Re^1$ the only parallels are given by the

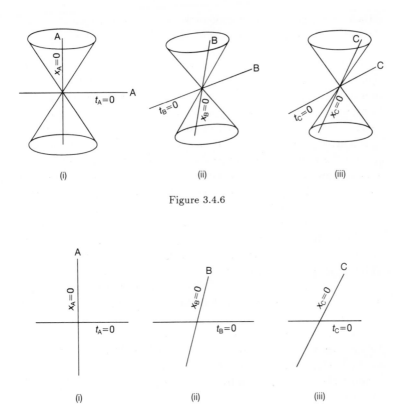

Figure 3.4.6

Figure 3.4.7

lines of absolute simultaneity, each of which picks out a particular $\Re^3$—the whole Euclidean space at a given absolute time—and which therefore leaves the $\Re^3$ component metrically amorphous. In Minkowski space by contrast the light cones provide two families of parallel lines in any $(1 + 1)$-dimensional subspace and so give access to affine geometry in these subspaces, and hence in the whole space. To go from affine geometry to full metric geometry we need some definition of congruence. Zeeman uses parallel projections. Provided there is more than one $(1 + 1)$-subspace parallel projections will only preserve one-one-ness if they preserve congruence (up to a scale factor). Robb and Winnie use orthogonality. One of Euclid's axioms had been that all right angles are equal; once we can identify some angles as being right angles, we can begin to impose a metric on affine geometry. The optical parallelogram is the key to the metric structure of Minkowski spacetime.

In Chapter 2,[17] we considered Einstein's assertion that the assignment of the value $\frac{1}{2}$ to ε in the Radar Rule was purely conventional, and claimed that although any value between 0 and 1 could be assigned to ε without being incompatible with the observational data, it was none the less reasonable to put ε equal to $\frac{1}{2}$ not just as a matter of convenience but as according with the symmetry of space. We now see a deeper rationale. The Robb rectangle correlates lines of simultaneity with world-lines at rest in a frame of reference, and does so in accordance with the Radar Rule with ε set at $\frac{1}{2}$. Instead of being an extra assumption we have to make in order to be able to assign dates to distant events, the Radar Rule, in its standard form, falls out of the Robb rectangle as the natural way of assigning dates and measuring durations and distances. "The one-way velocity of light" ceases to be a problem, granted the absolute approach, because instead of having to make a special, unverifiable assumption that the speed of light is the same on its return journey as it was on its outward journey, we find that it is a consequence of the light-cone structure of Minkowski spacetime. As we obtain a more stereoscopic view of the Special Theory, we see that what from one point of view appeared to be quite independent theses can be seen from another point of view to be deeply connected.

The argument of this section is reviewed in §3.6

---

[17] §2.8, pp.58-61.

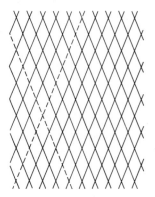

Figure 3.5.1 This is the same as Figure 3.2.1 where we saw two families of parallel lines generated by the light cones whose vertices are in the plane.

§3.5   Mackie's Argument for Absolutism

J.L. Mackie argues that the Special Theory is, despite appearances, committed to a Newtonian doctrine of Absolute Rest.[18] After drawing a number of valuable distinctions and making valid points, he invites us to consider the spacetime diagram constituted by a grid of lines representing two families of light rays, as in Figure 3.2.1, repeated here as Figure 3.5.1 above.

He then considers any point of intersection of light rays, and its light cone, and two photons emitted in opposite directions. There will be an event of the left-hand photon being at C, which is causally intermediate between its being emitted from O and arriving at A (see Figure 3.5.2).

He then argues by symmetry that there is exactly one corresponding point, D, along the line of the light ray going in the opposite direction. Granted this, we have, instead of the optical parallelogram of Robb, an "optical square", defining a unique event G, when the forward light cones of C and D intersect, and the world

---

[18] J.L.Mackie, "Three Steps Towards Absolutism", §4, in Richard Swinburne, ed., *Space, Time and Causality*, Dordrecht, 1983, pp.17-22.

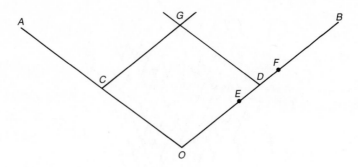

Figure 3.5.2 Mackie claims that there is just one event *D* corresponding to the arrival of the left-hand photon at *C*.

line *OG* will be at rest, absolutely at rest. Mackie anticipates the objection that our determination of the event *G* depends on our frame of reference, and parries it with anti-verificationist argument. "I do not claim," he says, "that there is any way of *determining* the point that corresponds to *C*; I say only that *there is* one. And nothing but the sort of extreme verificationism which I mentioned, but set aside, in Section I would rule out this claim as meaningless. Verificationism apart, there may well *be* things which we cannot discover or identify..."[19] But the objection is not a verificationist one. It is not that there is no way of determining the point, but that 'corresponds', like the other equivalence relations listed in Chapter 2,[20] is a triadic relation, and does not acquire definite sense until the sort of correspondence is specified. We need to know in this case what the axis of symmetry is, before we can even mean, let alone determine, the symmetrical correlate to *C*. In Figure 3.5.3 three different axes are suggested, which will make Mackie's points *E*, *D* and *F* correspond to *C*.

Correspondence does not have to be specified by means of a frame of reference. Dorling offers an alternative, in which corresponding events on the tracks of the two photons are specified in terms of wave crests. But that correspondence is not an absolute one. It is, in effect, simultaneity in that frame of reference in which the source

---

[19] pp.18-19

[20] §2.1, pp.28-29.

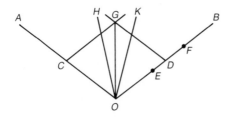

Figure 3.5.3 The three frames of reference in which OH, OG and OK are at rest will indicate E, D and F respectively as corresponding to C.

of the photons is at rest. Else, if the source is moving in the frame of reference, there will be Doppler shift in frequencies and wavelengths, and the correspondence measured by counting wave crests will not have any physical significance; if I have three shillings in my pocket, and you three dimes, it would be implausible to make out that we corresponded wealth-wise.

Mackie's mistake arises from not realising that 'corresponds' is a triadic relation. He supposed that the demand for a specification of the exact sort of sameness involved was merely fuelled by an illegitimate verificationism, and that once he had refuted that, nothing was left of the objection. But the objection, though no doubt sometimes couched in verificationist terms, is not really anything to do with how we might know what point corresponded to C, but rather with what is actually meant: it is based not on bad epistemology, but on good logic. Thus far Mackie's argument and the counter to it belong to the previous chapter: but in the light of the suprising way in which a metric emerges from Robb's approach, and the general way in which Mackie sets out the problem, we need to consider further whether it might not be possible to define an absolute sense of 'corresponds' by means of the Optical Parallelogram. Might it not be possible, we should ask ourselves, to construct an "Optical Rhomboid", or even an "Optical Square" which would determine an G uniquely?

It seems unlikely. The metric that Robb's construction generates is the standard metric of Minkowski space, which assigns zero length to every light ray. The Optical Parallelogram, which is essential for defining orthogonality and congruence, becomes degenerate when its diagonals are themselves light rays. The spacetime diagrams we draw to illustrate Robb's geometry are misleading in so far as they suggest symmetries that that geometry cannot define. As with Euclidean representations of topological truths, we need to be wary of reading more out of the diagrams than is really warranted. Although it would be rash to claim, without proof, that Robbian geometry could yield no natural canon of correspondence along light rays, Mackie has given us no reason to suppose that it can.

Mackie's argument fails. But it is an illuminating failure, and should warn us not to be too dogmatic about what is and is not possible when arguing about the Special Theory. As we shall see in §3.9 there are arguments from outside the Special Theory which indicate that there is some preferred canon of simultaneity, and hence a preferred frame of reference and world lines of absolute rest. And though it would be very surprising if there were arguments to this effect arising from within the Special Theory itself, it cannot be ruled out as absolutely impossible *a priori*.

§3.6 Critiques of the Absolute Approach

The Absolute Approach can be criticized. Not only is it forbiddingly abstract, but it is often associated with a reductionist philosophy that does scant justice to the fullness of scientific truth. It has been taken to support a relationist or some other reductionist account of time, and to argue against the invariance approach of the previous chapter. It seems to suggest that we can give an adequate account of time and space in terms of causality alone: granted one simple causal relation, $\ll$, we seem to be able to derive the whole of spacetime physics, thus warranting a reductive analysis of space and time to just the relation of causality. But it is not clear that cause is a more basic or better understood concept than those of temporal priority or distance. Normally the argument is from *post hoc* to *propter hoc* rather than *vice versa*. Cause is a difficult concept, being linked not only to time, but also to explanation.[21] Moreover, as with all relationist programmes, there is a modal difficulty. The relation A can causally influence B includes a modal term 'can', which takes us beyond the bounds of actuality. Not everything does, as a matter of fact, influence other events that it conceivably could. Moreover, to give a relationist account, we need not only a relation, $\ll$, but a relational structure $< X, \ll >$; that is, we need to say not only what the relation is, but what sort of entity it holds between. The entities are not ontologically simple events, but *possible* events, which are far from simple. In order to give an account of spacetime in terms of an absolute approach, we have to go beyond discoverable causes between existing events to potential causes between possible events, and it is far from clear that these either are more easily known or exist in a more straightforward way than the spacetime they were supposed to supplant. Moreover, concepts of continuity and dimensionality are needed, which cannot be derived from causality alone; and a fundamental distinction between timelike and spacelike separation is understood in the light of our previously established concepts of time and space.

For these reasons the reductionist programme is unattractive, and many thinkers have concluded that the absolute approach is unattractive too. But that does not follow. Not all absolute approaches need be reductionist. If instead of reducing spacetime to

---

[21] See *Space, Time and Causality*, Chapter 3, pp.37-41.

causality, we see it as the arena within which causal processes take place, the absolute approach shows the constraints on the structure and the metric of spacetime that this *rôle* imposes. At the very least, it shows that, granted a reasonable ontology and a satisfactory notion of causal influenceability, we can establish the Special Theory on quite different grounds from those normally adduced in its support, and that even if these are not claimed to be metaphysically prior, they still would constitute a constraint on the form that the theory could take, and explain some of its peculiar features.

Graham Nerlich makes cogent criticisms of some understandings of the absolute approach, and argues further against the use of the "Limit Principle".[22] The Limit Principle characterizes the light cone as the upper limit of the propagation of causal influence; it is the third approach outlined in §3.1 above. Nerlich argues that the possibility of tachyons—entities travelling faster than the speed of light—should not be excluded. The possibility of tachyons will be discussed in the next section, but certainly it should not be ruled out in advance by the way the Special Theory is presented. But since tachyons, as we shall see, cannot interact with entities going slower than light, it is possible to restate the limit principle so as to set an upper bound to only "subluminal entities"; if a body is capable of being at rest, then it can only go at speeds less than that of light.

Nerlich's main argument is that the "Invariance Principle" is more characteristic of the Special Theory than the Limit Principle. Essentially, it is about the value of different ways of characterizing the Special Theory. The attraction of the Limit Principle is epistemological: it starts from what we already know, and introduces us to the Special Theory as following from a perfectly intelligible limit to the speed with which causal influence can, as it happens, be propagated. But by itself it would not take us very far. It characterizes the interior of the light cone, the Absolute Future and the Absolute Past, but the main work of the absolute approach is done not by the interior of the light cone but by its surface. Whether we start with causal influenceability in the narrower sense or in the

---

[22] Graham Nerlich, "Special Relativity is not Based on Causality", *British Journal for the Philosophy of Science*, **33**, 1982, pp.361-388; see also Brent Mundy, "The Physical Content of Minkowski Geometry", §7, *British Journal for the Philosophy of Science*, **37**, 1986, pp.45-49.

wider sense or with light-connectability itself, we have to define some other relation, and pick out the light cone of each event as the characteristic topological feature of Minkowski spacetime. To this extent at least, Nerlich is right to insist that the Limit Principle is not, from a theoretical point of view, the most profound characterization of the Special Theory. It does not follow that the Invariance Principle, in either of the forms propounded by Nerlich, is more fundamental.[23] In one respect, certainly, ordering relations offer a better approach. They are asymmetric. There is a difference between $a \prec b$ and $b \prec a$, whereas with an equivalence relation, if $a \approx b$, then $b \approx a$. Most treatments of the Special Theory, like those of Newtonian mechanics, are completely time-reversible, and have great difficulty in accommodating the anisotropy of time, the fact that we do not think it is all the same whether time goes backwards or forwards. It is a merit of the absolute approach that it starts with an asymmetrical ordering relation, and so preserves the directedness of time throughout, and does not have to try to re-insert it at some late stage, for example by throwing away advanced potentials.[24]

The absolute approach is difficult to appreciate at first, but has great conceptual merits. It is difficult because it is abstract, and far removed from the phenomena physicists are familiar with. It was for this reason that we developed Einstein's approach first. Like Einstein's approach, it is based on a transitive relation, $\prec$, but one that is an ordering, rather than an equivalence, relation,

---

[23] But see below, §10.5, p.286.

[24] Although the relations $\prec$, $\ll$ and $\longmapsto$ are asymmetrical, they have a sort of symmetry none the less—there are no formal properties of $\prec$, which are not also possessed by $\succ$. This has led Mehlberg and Winnie to develop the absolute theory in terms of some symmetric relation, such as $a\gamma b$ where $a\gamma b$ if and only if $a \prec b \vee a \succ b$, which we can read as "a is causally connectable with b", or $a\lambda b$, where $a\lambda b$ if and only if $a \longmapsto b \vee b \longmapsto a$ which we can read as "a is light-connectable with b". But then it is still necessary to orient the spacetime, to say which of the two directions is future and which past. Nothing is gained by such a treatment. It is true and important that there are formal similarities between each of the relations employed in the absolute approach and its converse, but this is not the same as its being symmetrical, and so it is more illuminating to work with asymmetric relations throughout.

which, therefore is faithful to the ordered structure of our ordinary temporal experience. Granted that this is not only an ordering relation, but always a *partial*, and never a strict, ordering relation,[25] and satisfies some other conditions we should naturally expect the relation of cause and effect to satisfy in any sort of spatial and temporal manifold, we can derive not only the topological properties of the light cone, but the concepts of orthogonality and congruence, and hence most of the metrical properties of Minkowski spacetime. We can thus show that the metrical properties of Minkowski spacetime, and in particular setting $\varepsilon = \frac{1}{2}$ in Einstein's Radar Rule, are not in any way conventional or arbitrary, but flow from the basic topology induced by the partial ordering $\prec$.

§3.7 Tachyons

If we accept Nerlich's criticism of the absolute approach, and take the Lorentz transformation as the essential characteristic of the Special Theory rather than the speed of light's being the maximum attainable, we are led to speculate about the possibility of there being greater speeds. Such speeds can easily be imagined in a trivial sense. If I reflect a spot of light onto a wall by means of a mirror, I can, by rotating the mirror rapidly, cause the spot to move over the wall with a speed greater than that of light. Nor is there any logical inconsistency in our being able to communicate with extra-terrestrial beings outside our galaxy telepathically and instantaneously. We cannot rule out the possibility of superluminal speeds *a priori*, and need to consider their bearing on the Special Theory, even though at first sight it would seem that the argument of Chapter 1 would rule out any speed greater than that of light. But that argument was concerned with *universal* speeds, not with there being a *maximum* speed, and states that in any system subject to the conditions stated there can be only one universal speed, such that it is unaltered by being compounded with any other speed.[26] This indeed can be proved. It rules out the possibility of there being both the speed of light and an infinite speed of causal propagation, both playing the part of universal speeds. The

---

[25] That is, for every possible event there is another possible event it cannot influence or be influenced by.

[26] §1.2, p.7.

Michelson-Morley experiment showed that as far as electromagnetism was concerned the universal speed was the speed of light, and thus displaced Newtonian physics in which the universal speed was infinite. If, however, we concern ourselves not with universal speeds, but any propagation of causal influence with a speed greater than the speed of light, no inconsistency ensues, although we feel it to be profoundly unrelativistic, as indeed it is according to Einstein's formulation. Other formulations, however, can accommodate superluminal speeds, and we should not rule them out *a priori*.

Particles moving with a speed greater than that of light are called tachyons. They have space-like trajectories, along which the spacetime separation satisfies the condition $\sigma^2 > 0$, in contrast to ordinary material objects, whose rest mass is greater than zero (which have been called "tardyons", although this name is hardly ever used), whose trajectories are time-like, lying within the light cone and with a spacetime separation $\sigma^2 < 0$, or in terms of proper time, $\tau^2 > 0$.[27] The space-like character of the separation along the trajectory of a tachyon implies a spacelike four-momentum, so that $|p| > E$ and the rest mass m_0 is imaginary. But since the apparent mass

$$m = \frac{m_0}{\sqrt{(1 - v^2/c^2)}} \qquad (3.7.1)$$

and $v > c$, the apparent mass is real. Thus provided tachyons are never brought to rest, their imaginary mass is not inconsistent with the Special Theory. The expressions for the energy and the momentum of tachyons are

$$E = \frac{\mu c^2}{\sqrt{(v^2/c^2 - 1)}} \text{ and } |p| = \frac{\mu v}{\sqrt{(v^2/c^2 - 1)}} \quad (\mu \text{ real}), \quad (3.7.2)$$

so that $0 < E < \infty$ and $\mu c < |p| < \infty$. Thus tachyons speed up as they lose energy and have no energy at all when they are moving at infinite speed. Although more detailed studies of the theory of tachyons show that their behaviour would give rise to apparent causal anomalies, this still would not constitute a decisive physical argument against their existence, since no energy is transmitted.[28]

---

[27] See above, §1.5, p.18.

[28] G.Feinberg, *Physical Review*, **159**, 1967, pp.1089-1105; P.C.W.Davies, *The Physics of Time Asymmetry*, Surrey University Press, 1974,

No contradiction is implied by the existence of tachyons, and their existence cannot be completely ruled out on theoretical grounds, though there is at present no experimental evidence in its favour.

§3.8 Superluminal Velocities: The Bell Telephone

Conceptual arguments for there being superluminal velocities have been advanced. Some cases, such as the moving spot of light, are easy to accommodate: the spot is not really a *thing*; it has no independent existence, has no energy associated with it, and cannot carry a signal from one point to another. Other cases from quantum mechanics raise difficult metaphysical questions about the nature of quantum mechanical reality: properly understood, they pose no threat to the Special Theory, but we may be led by them, as also by considerations from theology, to view the Special Theory in a wider setting.

The collapse of the wave packet in quantum mechanics may be somewhat similar to the movement of the light spot. At any instant the probability of finding a particle at any particular place is given by the modulus squared of its wave function at that place. When we actually make a measurement on the particle, we find that it is at one point and at no other, and so we are inclined to to say that the wave function has instantaneously collapsed to that point. How this is to be interpreted depends on our philosophy of quantum mechanics. According to the Copenhagen interpretation, the wave function contains all that can be known about that particular particle, and so some instantaneous collapse appears to take place at the instant of measurement: according to Einstein's interpretation, however, the wave function gives only the probability distribution of an *ensemble* of similar particles, so that there is no need to speak of an instantaneous motion of anything. Probability distributions do not move at all. I may know that a colleague is either in his rooms in college or else in the Bodleian library, and so the corresponding probability distribution is concentrated on his being in these two areas. If I then telephone him in college, and he answers the phone, the probability distribution for his being in the library immediately vanishes, and that for being in his room increases to unity. But it would be absurd to say that one part of the probability distribution had instantaneously travelled from the library into the college, up the stairs and into his room. The instantaneous collapse of the wave function is not due, on this interpretation of quantum mechanics, to something's moving with a speed greater than that of light, because a wave function is only a probability, and a probability is not a thing at all.

In recent years there has been much discussion of the Bell inequalities, and, according to the Copenhagen interpretation, it has been suggested that the results of experiments made to see if they are violated show that "non-local" interactions can take place, implying the possibility of superluminal speeds. According to the realist interpretation, on the other hand, it is not surprising that the Bell inequalities are violated, since implicit in all the proofs of the inequalities is the assumption that a measurement on a system does not affect the result of a subsequent measurement, which is contrary to quantum mechanics.[29] It has even been suggested, on the former interpretation, that non-locality could be the basis of an "unstoppable and unjammable command-control-communication system at very high bit rates for use in submarine fleets".[30] If this were so, it would be of undoubted practical value.

Consider two particles, each of spin $\frac{1}{2}$ and bound together in a singlet state of spin zero, and suppose that the particles separate, and travel a long way from each other, the total spin of the whole system considered together must still be zero. According to quantum mechanics, if the measurement of the spin projection, along a certain direction, of one of the particles gives the result $+\frac{1}{2}$, then the measurement of the spin projection of the other particle in the same direction would give the result $-\frac{1}{2}$, and *vice versa* if the measurement of the spin projection, along a certain direction, of one of the particles gives the result $-\frac{1}{2}$, then the measurement of the spin projection of the other particle in the same direction gives the result $+\frac{1}{2}$. According to the Copenhagen interpretation the two particles, as they travel away from their common source, still constitute a single entity and do not contain any information about the spin directions of the constituent particles. It is only when the measurements are made that their projections are determined. Thus it looks as if a signal had to go instantaneously from

---

[29] T.A.Brody, "Where does the Bell Inequality Lead?", *Instituto Fisica Universidad Autonoma de Mexico*,1, 80-81.

[30] Quoted by N.D.Mermin, *Physics Today*, April 1985, p.38.

one to the other at the instant of measurement, to ensure their anti-correlation. This is described as a "non-local effect". Even after the separation the two particles still form one system, so that the measurement of one projection automatically determines the other.

Even if this account be true, it is difficult to see how information could be transmitted. Presumably the submarine would carry one of the particles and the control centre the other; then when a message was to be sent, the particle in the control centre would be forced into one state, whereupon a simultaneous measurement in the submarine would record the other state: with a sufficient number of particles, any message could be sent, instantaneously and with no possibility of interception.

But it will not work. In the first place it is not possible to force the particle at the control centre into a pre-chosen spin-projection. We can decide in what direction to measure spin, but we have no way of deciding in advance what the result of our measurement shall be. Although at the control centre we might choose to measure spin in the X direction, we could not know, until the measurement was made, whether the particle would be spin-up or spin-down in that direction. When we did know, we should also know the state of the particle in the submarine, but this knowledge would be useless for transmitting information. The crew of the submarine might have arranged previously to measure spin in the same direction, but all they could know would be the result of the measurement at the control centre, itself not under the control centre's control. What is under control is not the outcome of the measurement but the direction in which the spin is to be measured: we can control the questions we ask of nature, but not the answers nature gives. But this control is also useless for sending information. If the control centre alters the direction in which it measures spin, the submarine will never know: if it happens to align its measuring apparatus in the same direction, there will be an anti-correlation, which could be discovered suitably long after the event, but not until reports of the results obtained at the control centre had arrived by conventional methods. If it aligns its measuring apparatus in some other direction, the wave function of the particle will be forced into some eigenstate yielding an eigenvalue appropriate to the different question the submarine is posing.

The Copenhagen interpretation, though widely accepted by

physicists, has serious philosophical difficulties, and cannot be taken to be definitively established. Admittedly, other interpretations have their difficulties too, and we are far from understanding the reality behind quantum phenomena. Some principle of nonlocality may be called for; but if any superluminal speed is required, it is unlikely to be the speed of any *thing* or of any propagation of causal influence.

§3.9 Superluminal Velocities: Preferred Frames

This section develops further the argument of §2.9.

The Special Theory is concerned primarily with the movement of material objects and the propagation of causal influence. But Einstein's principle of equivalence has commonly been taken to mean that every inertial frame is equivalent to every other one, in the sense that there is nothing whatsoever to lead us to prefer any one to any other. And that, as we saw in §2.9, has far-reaching implications. For if we were able to accord some special significance, physical or metaphysical, to a canon of simultaneity, it would be enough to single out the corresponding frame of reference, even though no object or causal influence could travel instantaneously along it. If we believe that the collapse of the wave function happens independently of our coming to know about it, and that Schrödinger's cat either is alive or else is dead before we look at it, and cannot be a superposition of an alive wave function and a dead wave function, then we are committed to there being a difference between the future on the one hand and the present and the past on the other in much the same way as those philosophers who have claimed that they are modally different, and that the future is open whereas the past is unalterable and fixed. Equally, a theist who believes in God, and claims that God is omniscient and can have temporal knowledge of what goes on in the world, is committed to there being a preferred frame of reference too. In both cases there is a worldwide present moment, when either the wave function collapses into an eigenstate or when God knows events at different place to be happening concurrently, and so a preferred hyper-plane of simultaneity, and hence, more generally, a preferred frame of reference. The account of simultaneity given by the Special Theory is thus faced with a problem not only of quantum-mechanical events but of divine omniscience. If there is an ontological difference between future and past, then there is a real difference throughout the universe, giving us an absolute concept of simultaneity which is not dependent on the frame of reference, quite contrary to the teaching of the Special Theory, as commonly expounded.[31]

But in such arguments the real meaning of the Special Theory is being misunderstood. There is a confusion between what frames are equivalent to one another in respect of certain physical laws and what frames are equivalent to one another in all respects.[32] The best way of bringing this out is to consider the parallel case of classical mechanics. Classical mechanics, as we have seen, is relativistic not only with regard to displacement and re-orientation but with regard to velocity as well. There is no way, so far as the laws of Newtonian physics go, of telling whether a frame of reference is at absolute rest or moving with a uniform absolute velocity. Newton was quite clear on this point. But although he could not tell by means of Newtonian mechanics whether a frame of reference was at rest or not, it did not follow that it was a meaningless question, or one that he could not possibly have answered. He was inclined to think, although he could not be sure, that the centre of mass of the Solar System determined such a frame of reference, and he might have found a more certain answer in his theological researches—he might have found it revealed to him from his study of the Book of Daniel. So too if the result of the Michelson-Morley experiment had been positive, we should have been able to determine the velocity of the earth with respect to the aether, and the aether itself would have constituted such a frame of reference at absolute rest. Although Newtonian mechanics could not by itself single out any one of a set of inertial frames of references as being at absolute rest, Newtonian mechanics plus electromagnetic theory might have been able to. In the same way, although the Special

---

[31] Part of the argument put forward by H.Putnam, "Time and Physical Geometry", *Journal of Philosophy*, **64**, 1967, pp. 240-247; reprinted in H.Putnam, *Mathematics, Matter and Method, Philosophical Papers*, I, Cambridge, 1979, pp.198-205, and C.W.Rietdijk, "A Rigorous Proof of Determinism Derived from the Special Theory of Relativity", *Philosophy of Science*, **33**, 1966, 341-344, and "Special Relativity and Determinism", *Philosophy of Science*, **43**, 1976, pp. 598-609, depends on the absolute nature of modal distinctions.

[32] See above, §2.3, pp.36-37.

Theory cannot single out any one of a set of frames of reference that are equivalent to one another under the Lorentz transformation as being at absolute rest, there is no reason why the Special Theory plus some other theory or some other consideration should not do so.

This indeed happens in cosmology and in some versions of the General Theory. In cosmology the background radiation that echoes the Big Bang constitutes a cosmic frame of reference.[33] In the General Theory we sometimes distinguish a preferred frame of reference which we regard as being at rest, and we should have reason to do so if we adjoined to Special Relativity some realist version of quantum mechanics in which the collapse of the wave function is a real event, not just a change in the knowledge of an observer. Consider some interaction between a microphysical system on board a space rocket and the microphysical system of some star, say α-Centauri, which would take place as soon as the rocket arrived; a measuring instrument on the rocket interacting with some source on the star so as to register whether an electron emitted by it was spin-up in the direction of the earth's axis or spin-down. There is a definite, real date, not ascribed by us but inherent in the microphysical system, when the wave function of the electron collapses into either spin-up or spin-down along the direction of the earth's axis. We then ask how this date tallies with different frames of reference, say that of the earth and that of the rocket. It is unlikely that we could in practice tell, but it would no longer be in principle meaningless to ask the question. Let us consider the supposition that the collapse of the wave function is, in fact, simultaneous with our raising the question on earth. What follows? As far as the physics of the Special Theory is concerned, nothing follows. The event is outside the light cone in the Absolutely Elsewhere and Anywhen. It cannot affect, or be affected by, anything we do here and now. But we shall, in principle, be able to distinguish between frames of reference when we are concerned

---

[33] See G.F.Smoot *et al.*, *Physical Review Letters*, **39**, 1977, p.898; P.J.E. Peebles, *Physical Cosmology*, Princeton University Press, 1971; S.Weinberg, *Gravitation and Cosmology: Principles and Applications of the General Theory of Relativity*, Wiley, New York, 1972.

not with questions of physics,[34] but modal questions about time. If the collapse of the wave function is, in fact, simultaneous with our raising the question on earth, and if, according to the Radar Rule, the date assigned to the collapse of the wave function is the same in the frame of reference of the earth, then the earth's frame of reference is a preferred frame of reference, and is, really although not discoverably so by any electromagnetic phenomenon, at rest. And this will be no worse than what already we are familiar with in the General Theory, and what Newton envisaged for Newtonian mechanics.

The situation, however, may be a bit worse. There may be no preferred *inertial*[35] frame of reference. Again, consider the collapse of the wave function, not just on α-Centauri, but throughout the universe. There would be a tide of presentness, as the probabilistic distribution of the wave function collapses into actuality, into some one definite eigenvalue or another. This tide would constitute a 3-dimensional hyper-surface in spacetime, but need not be a hyper-plane. Only if it were a hyper-plane would it constitute the canon of simultaneity of some inertial frame of reference. Otherwise there would be no world-wide *inertial* frame of reference, but only different ones for different pairs of events. Instead of the flat spacetime of the Special Theory, we should be having to deal with a curved one in which the significant hyper-surfaces were not hyper-planes. But that, though not envisaged by Newton, is something we have been forced to countenance in the General Theory anyhow. It was a *desideratum* of the Special Theory that we should be able to find inertial frames of reference, not a *datum* that we always can.[36]

This conclusion runs counter to much current thinking which confuses science with the whole of knowledge. Science, and in particular physics, limits its aims and methods. Only certain sorts of proposition count as scientific laws or scientific observation: only

---

[34] It is not yet clear how widely Einstein's equivalence principle should range. Clearly, it covers all electromagnetic phenomena, but many physicists, argue that weak, strong and gravitational interactions also cannot be propagated with a speed greater than that of light. See J.Lévy-Leblond, "One More Derivation of the Lorentz Transformation", *American Journal of Physics*, Vol. 44, No.3, March 1976, p.271.

[35] See above, §2.5, pp.42-43.

[36] See above, §2.12, p.82.

certain sorts of observation or inference count as capable of establishing a scientific truth. Physics, above all other sciences, seeks
generality and therefore discounts many types of particularity as
irrelevant to its purposes. This concentration of aim and method
has resulted in far-reaching and deep insights, but is a limitation
none the less. The scientist, if he is wise, recognises that science
does not encompass the whole of knowledge, and that history for
example, although not science, is none the less knowledge. Similarly in his philosophy he needs to distinguish between questions
to which there is no scientific answer and questions to which there
is no conceivable answer. The latter may well be dubbed meaningless questions, not worth further discussion: the former are not
meaningless questions, although they are ones the scientist is not,
qua scientist, competent to answer, and so may be unwilling even
to discuss. But such unwillingness should not be thought a virtue:
the frontiers of science are not fixed, and questions that at one time
are unanswerable by science may later yield to scientific method.
For Newton whether there was an absolute frame of reference was
a theological question: for us it may be a quantum mechanical or
cosmological one as well.

§3.10 Conclusion

The absolute approach is unfamiliar to physicists. It seems remote
from experimental evidence, and forbiddingly *a priori*. Although
at first sight it seems to fit empiricist predilections for basing the
Special Theory on causality rather than on some abstract concept
of spacetime, it is no more successful in achieving that than relationist theories are generally. And although from an empiricist
point of view it is attractive to regard the speed of light as the upper
limit of propagation of causal influence, it gives less of an understanding of the special features of the Special Theory than than
that obtained from the standpoint of Lorentz invariance. Nevertheless, the absolute approach yields important insights. It reveals
the connexion between the topological and metrical properties of
Minkowski spacetime. It explains why the time-like and space-like
axes rotate in opposite directions with a change of relative velocity.
It justifies Einstein's setting ε equal to $\frac{1}{2}$. And being based on an
asymmetric transitive relation rather than a symmetric transitive
one, the absolute approach alone has built into it the anisotropy of
time.

Chapter 4
The Communication Argument

§4.1 From Radar Rule to Lorentz Transformation

This chapter presupposes ch. 2, esp., §§2.3 and 2.4.

Einstein proposed the Radar Rule as a way of ascribing dates and distances to distant events.[1] It is not obvious that the way the Radar Rule assigns dates and distances is the same as the way the Lorentz transformation does, but in fact it is, and Whitrow and Milne developed an elegant and thought-provoking derivation of the Lorentz transformation from the Radar Rule, granted certain other, seemingly plausible, assumptions. One of the extra assumptions is that electromagnetic radiation not only is reflected back, so as to locate distant events in a given frame of reference, but can be received and understood, so as to be a means of communication between different observers representing different inertial frames of reference, whereby their different ways of referring to dates, places and events, can be harmonized. We have, that is, in addition to the Radar Rule a "Radio Rule", which should be subject to certain symmetry conditions that express philosophically defensible ideals of parity of epistemological esteem between observers. If the Radar and the Radio Rules are to be coherent with each other, then the way different observers refer to dates, places and events, must transform according to the Lorentz transformation. Since the key to the derivation is that observers can communicate with one another as well as about distant objects and events, it should be seen as a particular type of communication argument.[2]

---

[1] A. Einstein, "Zur Electrodynamik bewegter Körper", *Annalen der Physik*, **17**, 1905, pp. 891ff., §1, tr. in G. Stephenson and C.W. Kilmister, *Special Theory of Relativity*, Oxford, 1970, pp. 188-191. See above, §2.8, p.59.

[2] Communication arguments are important in natural, as well as moral, philosophy. See *Space, Time and Causality*, Ch. VII, pp. 111–113, and *Treatise on Time and Space*, §8, pp. 45–47.

Most expositions of the communication argument start from very general assumptions, and proceed with the aid of some sophisticated mathematics to derive the Lorentz transformation in canonical form.[3] Although great care is taken to list all the assumptions, it is often difficult, on account of their generality, to assess exactly what each involves, and much of the labour seems to be devoted to producing a rather small mouse. We shall, therefore, depart from the normal mode of derivation, and work backwards rather than forwards, deriving the Lorentz transformation from fairly strong and specific assumptions, and only subsequently justifying them by reference to the more general ones. This is not only helpful as regards exposition, but illuminating philosophically. Much argument, in physics as well as elsewhere, proceeds not by deduction from axioms antecedently articulated, but by first formulating assumptions implicitly made use of in arguments already accepted and subsequently justifying those assumptions by appeal to more general considerations.

§4.2 The Derivation

Inertial frames of reference can be thought of in much the same way as families of Indo-European languages, Romance, Teutonic, Slavonic etc.[4] Each such family includes a variety of different languages, French, Italian, Spanish, etc., corresponding to particular co-ordinate systems, and can be represented by any one of them. If we want to transform from one inertial frame to another, we can simplify the working by choosing co-ordinate systems in which the direction of uniform relative motion is along the X-axis, and

---

[3] G.J. Whitrow, *The Natural Philosophy of Time*, Edinburgh, 1961, ch. II §8, pp.171–3 and ch. IV §§3–4, 2nd ed. Oxford, 1980, ch.5, esp. §§5.2-5.4, pp.230–253. E.A. Milne, *Modern Cosmology and the Christian Idea of God*, Oxford, 1952, ch. III and IV; and *Kinematic Relativity*, Oxford, 1948, ch. II, esp. §24. The exposition in this chapter is heavily indebted to C.W. Kilmister, *Special Theory of Relativity*, Oxford, 1970, ch.2, pp.14–19, and to G. Stephenson and C.W. Kilmister, *Special Relativity for Physicists*, London, 1958, ch.1 §7, pp.16–19 and C.W. Kilmister, *The Environment in Modern Physics*, London, 1965, ch.4, pp.46–53.

[4] See above, §1.5 and §2.7.

the temporal origin is set at the date when the spatial origins coincided. As in Chapter 2,[5] we shall suppose that French and German are in this sense suitable representatives of the Romance and Teutonic families, and work out what dictionaries they must use if they both use the Radar Rule and they each acknowledge the legitimacy and validity of the other language. The device of using different languages is difficult to extend into mathematical formulae. Normally physicists use dashes or primes to distinguish (x', y', z', t') from (x, y, z, t), but these are easily missed. The reader may find it helpful to use different-coloured biro in his working, but for printing we use different fonts of type.

A Frenchman, Monsieur Sanserif, a suitable representative of one inertial frame, is carrying on a conversation by wireless telephone with a German, Herr Muth, who is a correspondingly suitable representative of another inertial frame, moving at a uniform velocity with respect to the first.

Monsieur Sanserif écrit en français en sanserif, et pour les mathématiques en majuscules and **Herr Muth schreibt sehr mutig auf Deutsch.** How are they to translate dates, places and events? Not only *Venise* into **Venedig** and *Vienne* into **Wien,** but generally (X, Y, Z, T) into $(\mathbf{x}, \mathbf{y}, \mathbf{z}, \mathbf{t})$ and *vice versa?* To begin with,[6] we assume that they have a common origin, *i.e.*

$$(0, 0, 0, 0) = (\mathbf{0}, \mathbf{0}, \mathbf{0}, \mathbf{0}), \qquad (4.2.1)$$

which means that at time $T = \mathbf{t} = 0$, they are both in the same place, and we assume also that each of them is standing and remains standing at the origin of his co-ordinate system,[7] and that each reckons that the other is moving away from him with a uniform velocity with the same absolute magnitude; *i.e. M. Sanserif pense que Herr Muth s'éloigne avec vélocité V* and **Herr Muth glaubt, dass M. Sanserif mit Geschwindigkeit v zurückweicht** where

$$|\mathbf{v}| = |-V|. \qquad (4.2.2)$$

---

[5] §§2.5, 2.6.

[6] See below, §4.4, p.139.

[7] See above, § 2.5, pp.44-45.

Figure 4.2.1 **Herr Muth** talks with *Monsieur Sansserif* (modified from *Punch*, Vol. 147, December 2, 1914, with permission).

They both agree that the speed of light is finite, the same in all directions and the same for each of them; that is, $|C| = |c|$, which we shall therefore write c. They both use spacetime diagrams, though they picture them differently.

Suppose M. Sanserif sends a message to Herr Muth at time T_1, and that he calls the time the time it arrives at Herr Muth T_2, and at that time he reckons that Herr Muth is at a distance L from him.

Then since Herr Muth is moving at a constant velocity V, M. Sanserif deduces that

$$L = V T_2. \qquad (4.2.3)$$

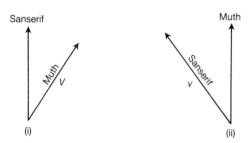

Figure 4.2.2   Spacetime diagrams for (i) M. Sanserif and (ii) Herr Muth.

The signal sent by M. Sanserif travels at velocity c, and so

$$(T_2 - T_1)c = L. \qquad (4.2.4)$$

Hence

$$T_1 = (1 - \frac{V}{c})T_2. \qquad (4.2.5)$$

Immediately Herr Muth receives the message, he sends a message back to M. Sanserif. Since they have previously arranged that he should do this, M. Sanserif can expect the message at a time T_3. If the message travels with the same velocity in both directions,

$$T_3 = T_2 + \frac{L}{C} = \left(1 + \frac{V}{C}\right)T_2. \qquad (4.2.6)$$

So that

$$T_2 = \frac{1}{2}(T_1 + T_3) \qquad \text{Radar Rule} \qquad (4.2.7)$$

and

$$T_1 = \frac{1 - V/C}{1 + V/C}T_3. \qquad (4.2.8)$$

Now let us suppose that each receives some signals from the other's clock: perhaps each sends out pips every hour, as the BBC does. We should not assume that each thinks that the other's clock is telling the same time as his own one. Rather, he may suppose that the other's clock is running uniformly slow (or uniformly fast) by a factor that could conceivably, and in fact does, depend on v. Thus if t denotes the time observed by Herr Muth on his clock, M. Sanserif supposes $t_2 = kT_1$ and Herr Muth supposes $T_3 = kt_2$.

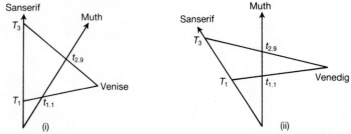

Figure 4.2.3 Spacetime diagrams showing (i) how M. Sanserif calculates, (ii) how Herr Muth calculates.

This assumes that the factor k is the same for both of them. In that case

$$T_3 = k^2 T_1 \qquad (4.2.9)$$

$$\text{and so} \quad k^2 = \frac{1 + v/c}{1 - v/c}. \qquad (4.2.10)$$

We note that although $\mathbf{v} = -V$, there is a compensating change of sign in c, so that $\mathbf{v}/c = V/c$. We therefore need not specify which language is being used, and write in either case v/c. The change of sign between V and $\mathbf{v}$ is not always insignificant, and needs careful watching, but in this case we can write

$$k = \sqrt{\frac{1 + V/c}{1 - V/c}} = \sqrt{\frac{1 + v/c}{1 - v/c}} = \sqrt{\frac{1 + \mathbf{v}/c}{1 - \mathbf{v}/c}}. \qquad (4.2.11)$$

(We may wonder what it would mean to replace v by $-v$. The answer is that then the two inertial frames of reference would be approaching each other, so that their origins, $(0, 0, 0, 0)$ and $(\mathbf{0}, \mathbf{0}, \mathbf{0}, \mathbf{0})$, that is the time when they both are in the same place, would be in the future, T_1 and $\mathbf{t}_2$ would be negative, and k would need to be less than unity in order to have the date of receiving a message after the date of transmission.)

It is convenient at this stage to calculate $k + \frac{1}{k}$ and $k - \frac{1}{k}$.

$$k + \frac{1}{k} = \sqrt{\frac{1 + v/c}{1 - v/c}} + \sqrt{\frac{1 - v/c}{1 + v/c}} = \frac{1 + v/c + 1 - v/c}{\sqrt{1 - v^2/c^2}} = 2\gamma,$$
$$(4.2.12)$$

where γ is defined, as in §1.3, p.13, and §2.6, eq.2.6.4, p.49, as $\frac{1}{\sqrt{1 - v^2/c^2}}$. A similar calculation shows that $k - \frac{1}{k} = 2\gamma v/c$.

Having established the value of k, we now consider a particular event, say the ringing of the bell in the Campanile in St. Mark's Square in Venice on a particular day, and see how M. Sanserif and Herr Muth will date and locate it. Let us assume that they are in line with one another, so that the spacetime diagram is like that in Figure 4.2.3, with the bell ringing when M. Sanserif's signal arrives.

Note that T_2 is now the date M. Sanserif ascribes to the arrival of his radar pulse at Venice, *not* the date it reaches Herr Muth. To make the sequence of events clearer, Herr Muth's dates have been called $t_{1.1}$ and $t_{2.9}$, instead of t_1 and t_3. *M. Sanserif pense que s'il envoie un message à temps T_1, il passera Herr Muth à son temps à lui $t_{1.1}$, et qu'il arrivera à Venise où il sera immédiatement rélé, té et passera de nouveaux Herr Muth à son temps à lui $t_{2.9}$, pour être de retour à temps T_3. La règle Radar dit:*

$$T_2 = \frac{1}{2}(T_3 + T_1)$$
$$L_2 = \frac{1}{2}c(T_3 - T_1),$$
(4.2.13)

où L_2 est la distance entre M. Sanserif et Venise. Il s'ensuit que

$$T_3 = T_2 + L_2/c$$
$$T_1 = T_2 - L_2/c.$$
(4.2.14)

Herr Muth glaubt, dass, wenn eine Botschaft nach Venedig an ihm zum Zeitpunkt $t_{1.1}$ vorbeigeht, in Venedig zuruckgeworfen wird und zum Zeitpunkt $t_{2.9}$ zuruckkommt, dann, nach der Radarregel

$$t_2 = \frac{1}{2}(t_{2.9} + t_{1.1})$$
$$l_2 = \frac{1}{2}c(t_{2.9} - t_{1.1}),$$
(4.2.15)

wo l_2 die Entfernung zwischen seinem deutschen Heimatsort und Venedig ist. Es folgt, dass ...

$$t_{2.9} = t_2 + l_2/c$$
$$t_{1.1} = t_2 - l_2/c.$$
(4.2.16)

But M. Sanserif and Herr Muth both agree that

$$T_3 = k\mathbf{t}_{2.9}$$
$$\mathbf{t}_{1.1} = kT_1,$$

(4.2.17)

where, as before,

$$k = \sqrt{\frac{1 + v/c}{1 - v/c}}.$$

(4.2.18)

Il s'ensuit que

$$T_2 + L_2/c = T_3 = k\mathbf{t}_{2.9} = k(\mathbf{t}_2 + \mathbf{l}_2/c)$$
$$T_2 - L_2/c = T_1 = \mathbf{t}_{1.1}/k = (\mathbf{t}_2 - \mathbf{l}_2/c)/k$$
$$\text{So} \quad 2T_2 = (k + 1/k)\mathbf{t}_2 + (k - 1/k)\mathbf{l}_2/c = 2\gamma\mathbf{t}_2 + 2\gamma(v/c^2)\mathbf{l}_2$$
$$\textit{i.e.} \quad T_2 = \gamma(\mathbf{t}_2 + (v/c^2)\mathbf{l}_2).$$

(4.2.19)

Et

$$2L_2/c = (k - 1/k)\mathbf{t}_2 + (k - 1/k)\mathbf{l}_2/c = 2\gamma(v/c)\mathbf{t}_2 + 2\gamma\mathbf{l}_2/c$$
$$L_2 = \gamma(\mathbf{l}_2 + v\mathbf{t}_2).$$

(4.2.20)

By an exactly similar chain of reasoning, **es folgt, dass**

$$\mathbf{t}_2 - \mathbf{l}_2/c = \mathbf{t}_{1.1} = kT_1 = k(T_2 - L_2/c)$$
$$\mathbf{t}_2 + \mathbf{l}_2/c = \mathbf{t}_{2.9} = T_1/k = (1/k)(T_2 + L_2/c)$$
$$2\mathbf{t}_2 = (1 + 1/k)T_2 + (1/k - k)L_2$$
$$\mathbf{t}_2 = \gamma(T_2 - (v/c^2)L_2).$$

(4.2.21)

Und

$$\mathbf{l}_2 = \gamma(L_2 - vT_2).$$

(4.2.22)

These rules for translating from German into French and *vice versa* are exactly the same, except for the sign of v. V and $\mathbf{v}$ are of opposite signs, and it makes a difference whether we view the relative uniform velocity from M. Sanserif's standpoint, who reckons Herr Muth is moving away in the $+X$-direction which he calls V, or from Herr Muth's standpoint, who reckons M. Sanserif is moving away with uniform velocity $\mathbf{v}$, which is in the $-X$-direction so that $\mathbf{v} = -V$. It depends, therefore, on whether we use V or $\mathbf{v}$, what sign we should have. We can construe this as a choice. when we want to translate Herr Muth's references into French, between **ein deutsch-französisches Lexikon** and *un dictionnaire allemand-français* and between **ein französisch-deutsches Lexikon** and *un français-allemand dictionnaire* when we want to translate M. Sanserif's references into German.

The first version is *un dictionnaire allemand-français*. It translates German into French, and is itself written in French (and so is in terms of the French V rather than the German $\mathbf{v}$). Written out in full and expressed in a more general form, replacing T_2 and L_2 by T and X, and $\mathbf{t}_2$ and $\mathbf{l}_2$ by $\mathbf{t}$ and $\mathbf{x}$, it is

$$\begin{aligned} \mathbf{x} &= \gamma(X - VT) \\ \mathbf{y} &= Y \\ \mathbf{z} &= Z \\ \mathbf{t} &= \gamma(T - VX/c^2). \end{aligned} \tag{4.2.23}$$

The second version is *un dictionnaire français-allemand*. It translates French into German, and again is itself written in French (and so is in terms of the French V rather than the German $\mathbf{v}$). Written out in full, and in general form, it is

$$\begin{aligned} X &= \gamma(\mathbf{x} + V\mathbf{t}) \\ Y &= \mathbf{y} \\ Z &= \mathbf{z} \\ T &= \gamma(\mathbf{t} + V\mathbf{x}/c^2). \end{aligned} \tag{4.2.24}$$

If Herr Muth had insisted on using **ein französisch-deutsches Lexikon,** he would have needed to replace the French V by the German $-\mathbf{v}$, in which case he would have

$$\begin{aligned} X &= \gamma(\mathbf{x} - \mathbf{v}\mathbf{t}) \\ Y &= \mathbf{y} \\ Z &= \mathbf{z} \\ T &= \gamma(\mathbf{t} - \mathbf{v}\mathbf{x}/c^2). \end{aligned} \tag{4.2.25}$$

For the sake of completeness, we should note also **ein deutsch-französisches Lexikon,** in which Herr Muth can translate his references in terms of the velocity $\mathbf{v}$ he ascribes to M. Sanserif, once again with a change of sign. The four transformations are tabulated below:

Table 4.2.1   **The Four Dictionaries**

dictionnaire *allemand–français*	*dictionnaire* *français–allemand*
$\mathbf{x} = \dfrac{X + VT}{\sqrt{1 - v^2/c^2}}$	$X = \dfrac{\mathbf{x} - V\mathbf{t}}{\sqrt{1 - v^2/c^2}}$
$\mathbf{x} = Y$	$Y = \mathbf{y}$
$\mathbf{z} = Z$	$Z = \mathbf{z}$
$\mathbf{t} = \dfrac{T + VX/c^2}{\sqrt{1 - v^2/c^2}}$	$T = \dfrac{\mathbf{t} - V\mathbf{x}/c^2}{\sqrt{1 - v^2/c^2}}$
deutsch–französisches **Lexicon**	**französisches–deutsch** **Lexicon**
$\mathbf{x} = \dfrac{X - \mathbf{v}T}{\sqrt{1 - v^2/c^2}}$	$X = \dfrac{\mathbf{x} + \mathbf{v}\mathbf{t}}{\sqrt{1 - v^2/c^2}}$
$\mathbf{y} = Y$	$Y = \mathbf{y}$
$\mathbf{z} = Z$	$Z = \mathbf{z}$
$\mathbf{t} = \dfrac{T - \mathbf{v}X/c^2}{\sqrt{1 - v^2/c^2}}$	$T = \dfrac{\mathbf{t} + \mathbf{v}\mathbf{x}/c^2}{\sqrt{1 - v^2/c^2}}$

§4.3 The Radar Rule Justified

The derivation given in the previous section is fairly simple but
depends on assumptions which may not be readily granted. It assumes that the co-ordinate systems can be chosen so as to have the
relative velocity along a common axis, and so as to have a common
origin; that the methods of measurement are the same for both observers in their different frames of reference, and in particular that
the Radar Rule in its standard form is unquestionably correct; and
that it is right to suppose that the date of reception by one observer
of a signal from the other is proportional to the duration that has
elapsed since their being together until the date of its transmission,
and that the constant is the same in both directions. Each of these
assumptions may reasonably be questioned.

Arguments about the Special Theory often seem to cheat by taking specially simple co-ordinate systems, in this case with a common origin and the relative motion along the X-axis. Why, in the terminology of §2.3, should we choose French and German, conveniently facing each other across the Rhine, as the representatives of the Romance and Teutonic inertial frames, instead of, say, Latin and Danish, or Rumanian and Dutch? The answer is that we *can* have co-ordinate systems that are displaced along a common axis, or whose axes are not in the same direction, but that the transformations involving displacement or re-orientation, though clumsy, do not raise points of significance as regards the Lorentz transformation. Rather than consider the clumsy translation from Italian to Gothic and *vice versa*, we translate clumsily, but unproblematically, from Italian to Latin (which might involve a displacement of origin), and from Latin to French (which might involve another displacement of origin), and from Gothic to which might involve a re-orientation of axes), and from Danish to German (which might involve just a simple displacement of origin again), so as to concentrate our attention on the crucial problem, that of transforming from one co-ordinate system to another moving with uniform velocity with respect to it.

The assumption that M. Sanserif and Herr Muth should both use the Radar Rule is open to challenge. They are essentially distant from each other, and cannot compare lengths and durations directly. Their only means of communication is by wireless, each noting the time he transmits signals to, and receives signals from, the other. Their measure of distance, therefore, must be essentially timelike, like our light-years. It would be perfectly reasonable for them to adopt something like the second half of the Radar Rule

$$l = \frac{1}{2}c(t_3 - t_1) \qquad (4.3.1)$$

by agreement, as *defining* what each meant by distance. Instead of talking about 186,000 miles, where a mile is 1,760 yards, M. Sanserif and Herr Muth could agree on their wireless telephone to talk about "seconds of distance" along with "seconds of arc" and "seconds of time", meaning by a second of distance the distance away an object must be if it took two seconds of time to receive back a radar pulse from it. But although it would be reasonable to adopt some such rule for measuring distance, it is not obvious that

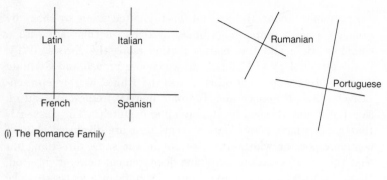

(i) The Romance Family

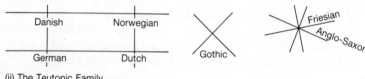

(ii) The Teutonic Family

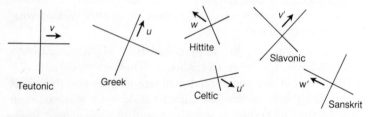

(iii) The Indo-European Families

Figure 4.3.1 Indo-European Language Groups. The Teutonic Family
is moving with velocity V to the right. It so happens that French
and German have their X-axis along the line of relative motion,
and were at one time coincident, and took the date of coincidence
as their temporal origin. So we take those two languages (*i.e.* co-
ordinate systems) as convenient representatives of the two families
(*i.e.* inertial frames).

the rule to be adopted should be this one. All we can insist on at
the outset is that in ascribing dates and distances to distant events
M. Sanserif and Herr Muth should agree that if e_2 is the distant
event of a radar signal, transmitted as event e_1 at date t_1, being

reflected so as to return as event e_3 at date t_3, then

$$t_2 = f(t_3, t_1)$$
$$l_2 = g(t_3, t_1). \tag{4.3.2}$$

where f and g are functions of t_3 and t_1. Yet, as we shall now show, even from this little the general form of the Radar Rule follows, and granted very modest additional premisses about the nature of distance and distant events, the specific Radar Rule will follow. The general form of the Radar Rule follows from their agreeing that dating systems are arbitrary, both as to origin and as to unit. It should not matter, as far as physics is concerned, whether we reckon our dates from the birth of Christ or the First Olympiad, or whether our units are years, quadrennia, hours, or nanoseconds — indeed, we have taken this for granted in working with frames of reference, that is to say, whole equivalence classes of co-ordinate systems, rather than some particular co-ordinate system. Let us consider the change of origin as it bears upon distance first. It is convenient to write $g(t_3, t_1)$ as a function of $(t_3 + t_1)/2$, $(t_3 - t_1)/2$, and say that

$$l_2 = \psi\left(\frac{t_3 + t_1}{2}, \frac{t_3 - t_1}{2}\right). \tag{4.3.3}$$

It is clear that a change of temporal origin should not affect the distance, so that if we add a constant A to all our dates, l_2 should be the same. That is,

$$\psi\left(\frac{t_3 + t_1}{2}, \frac{t_3 - t_1}{2}\right) = \psi\left(\frac{t_3 + t_1 + A}{2}, \frac{t_3 - t_1}{2}\right) \tag{4.3.4}$$

for all A. Hence ψ can depend only on $(t_3 - t_1)/2$, not on $(t_3 + t_1)/2$.

It is also very natural to suppose that if the unit of time is altered, the unit of length should be altered proportionately, if we are measuring length in essentially temporal terms. If we measure time in years rather than seconds, then we should reckon to be measuring distance in years of distance instead of seconds of distance, and to have the unit of distance, like the unit of time, increased by a factor of $60 \times 60 \times 24 \times 365$. In that case the function must be of the form

$$l_2 = C(t_3 - t_1), \tag{4.3.5}$$

where C is some constant. If we write $C = \frac{1}{2}c$, we have

$$l_2 = \frac{1}{2}c(t_3 - t_1), \tag{4.3.6}$$

although further argument is needed to identify this c with the speed of light.

The argument for the first half of the Radar rule is more complicated. Once again, it is convenient to replace $t_2 = f(t_3, t_1)$ by

$$t_2 = \phi\left(\frac{t_3 + t_1}{2}, \frac{t_3 - t_1}{2}\right). \tag{4.3.7}$$

We then argue that for any constant A (which might be the date of the First Olympiad, 778B.C.)

$$t_2 + A = \phi\left(\frac{t_3 + t_1}{2} + A, \frac{t_3 - t_1}{2}\right). \tag{4.3.8}$$

Hence, in particular, taking $A = -\frac{1}{2}(t_3 + t_1)$

$$t_2 = \frac{t_3 + t_1}{2} + \phi\left(0, \frac{t_3 - t_1}{2}\right). \tag{4.3.9}$$

If this is also to be indifferent as to unit,

$$\begin{aligned}
Bt_2 &= \phi\left(B\frac{t_3 + t_1}{2}, B\frac{t_3 - t_1}{2}\right) \\
&= B\frac{t_3 + t_1}{2} + \phi\left(0, B\frac{t_3 - t_1}{2}\right).
\end{aligned} \tag{4.3.10}$$

If this is to hold generally, $\phi(0, B(t_3 - t_1)/2)$ must be of the form $D \times \frac{1}{2}(t_3 - t_1)$ where D is a constant. So

$$t_2 = \frac{t_3 + t_1}{2} + D \times \frac{t_3 - t_1}{2}. \tag{4.3.11}$$

If we set $\varepsilon = (1 - D)/2$,

$$t_2 = \varepsilon t_1 + (1 - \varepsilon)t_3, \tag{4.3.12}$$

which is the general form of Einstein's Radar Rule.

To go further, and determine the precise value of ε, M. Sanserif and Herr Muth need to form a view about distant events, and their relation to the observer. It is clear, first, that the date and distance ascribed by the general rule

$$t_2 = \varepsilon t_1 + (1 - \varepsilon)t_3$$
$$l_2 = \frac{c}{2}(t_3 - t_1) \tag{4.3.13}$$

are intended to be assigned to one and the same event, e_2, which is reckoned to have occurred at date t_2 and in a place that is l_2 away. It is reasonable then to think of a one-way distance between events as analogous to the temporal duration between them. In assigning the date t_2 to e_2, we are assigning a temporal duration of $t_2 - t_1$, the time taken for the signal, whose transmission at time t_1 was event e_1, to arrive at its distant destination. The one-way distance must depend on e_2 and e_1, and since these are referred to, fundamentally, in temporal terms, on t_2 and t_1. And by the same reasoning as for l_2, the one-way distance function must depend only on the difference, $t_2 - t_1$, and be proportional to it.

It is also reasonable for each observer to think of his own frame of reference as being at rest, in which case he will think that e_3, the arrival of the reflected radar signal, occurs in the same place as e_1, so that the return distance from e_2 to e_3 will be the same as the outward distance from e_1 to e_2, though in the opposite direction. If they are to have the same absolute magnitude, $t_3 - t_2$ must be the same as $t_2 - t_1$, and so $t_2 = \frac{1}{2}(t_3 + t_1)$ and $\varepsilon = \frac{1}{2}$. The assumptions made are still vulnerable. It may be questioned why each observer should assume that he is at rest, and it may be complained that we are, in effect, stipulating that the velocity of light shall be the same on the return journey as on the outward one. The first objection arises from a misunderstanding. Our talk of observers is only a *façon de parler*, a vivid way of characterizing a frame of reference.[8] Real observers can move: but a frame of reference cannot move with respect to itself. It would be perfectly possible for a real M. Sanserif to choose a frame of reference—say that of an astronaut—in which he was moving, in which case outward and return journeys would be different: but if we choose a particular frame of reference, then

---

[8] See above, §2.5, pp.44-45.

in that frame of reference the frame of reference itself is necessarily at rest, and outward and return distances ought to be the same.

The velocity of light enters the argument in the constant of proportionality between distance and time. It would be logically possible to have it different, depending either on the velocity of what it was being reflected from, or more fundamentally, on the direction in which the radar pulse was sent. Instead of having the one-way velocity always $c(t_2 - t_1)$, we might have it $c'(t_2 - t_1)$ and $c''(t_3 - t_2)$. The former possibility, though intuitively attractive to us, brought up on Newtonian mechanics, would be more difficult to accept if the speed of light were the only available measure of distance that M. Sanserif and Herr Muth could both adopt. Much more attention, however, has been given to the possibility that there is an inherent difference in the speed of light in different directions. Such a difference cannot be conclusively ruled out by experiment alone. It was a recognition of this truth that led Einstein, having claimed that the general form of the Radar Rule must be

$$t_2 = \varepsilon t_1 + (1 - \varepsilon)t_3$$
$$l_2 = \frac{c}{2}(t_3 - t_1), \tag{4.3.14}$$

to set ε equal to $\frac{1}{2}$ "by convention", but that, as we saw in Chapter 2, is to misuse the word 'convention', since there are good general reasons for setting ε equal to $\frac{1}{2}$.[9] These are represented in Whitrow's presentation by his postulate of homogeneity[10] and by Kilmister in his assumption 5.[11] These can be controverted without evident inconsistency. The difficulty about the general approach of Chapter 2, within which we have been considering the Radar Rule, is that there is always room for dispute over the exact respects in which sameness obtains. If, however, we adopt the causal approach of Chapter 3, the case is altered. The Robb rectangle correlates lines of simultaneity with world-lines at rest in a frame of reference, and does so in accordance with the Radar Rule with ε set at $\frac{1}{2}$.

[9] See above, §2.8, pp.59-61.

[10] G.J. Whitrow, *The Natural Philosophy of Time*, Edinburgh, 1961, Axiom V, p.189, 2nd ed., Oxford, 1980, p.235.

[11] C.W. Kilmister, *The Environment in Modern Physics*, London, 1965, p.48.

Instead of being an extra assumption we have to make in order to be able to assign dates to distant events, the Radar Rule, in its standard form, falls out of the Robb rectangle as the natural way of assigning dates and measuring durations and distances. "The one-way velocity of light" ceases to be a problem, granted the absolute approach, because instead of having to make a special, unverifiable assumption that the speed of light is the same on its return journey as it was on its outward journey, we find that it is a consequence of the light-cone structure of Minkowski spacetime. As we obtain a more stereoscopic view of the Special Theory, we see that what from one point of view appeared to be quite independent theses can be seen from another point of view to be deeply connected.

§4.4   Massaging the Messages

In the simple version of §4.2, it was argued that the time of reception by one observer of a signal transmitted by the other should be proportional to the date of transmission: $t = KT$ when M. Sanserif sends a signal to Herr Muth, and $T = kt$, when Herr Muth sends a signal to M. Sanserif. The argument relied on the fact that they were moving at a uniform velocity with respect to one another, and shared a common origin. But how do they know that they are moving at a uniform velocity, and that $v = -V$, and above all that $k = K$? Each can establish that the other is moving at a uniform velocity by simple use of the Radar Rule on a succession of different occasions. It would then be possible to extrapolate backwards or forwards to the time of meeting, and adopt that as the temporal origin. (This would be equivalent to translating from, say, Portuguese into French, or Gothic into German.) Provided they both have good time-keepers, they might reasonably hope to agree how far they were apart, when they had been, or would be, together, and hence what their velocity must be. But we cannot take good time-keeping for granted. After all, it is one of the chief consequences of the Lorentz transformation that clocks run slow in moving frames. The Milne and Whitrow derivation makes no assumption of good time-keeping. To start with, each observer has a clock whose only guaranteed property is that it never stops and never goes backwards: in mathematical terms it represents a strictly monotonic function into the real numbers. Since, however, each can communicate with the other, correlations between different clocks can be established. M. Sanserif can send signals which say that they were transmitted at T_1, and Herr Muth can note that he receives them at $\mathbf{t}_{1.1}$ by his clock; and, conversely, Herr Muth can send signals which say that they were transmitted at $\mathbf{t}_{2.9}$, and M. Sanserif can note that he receives them at T_3. If they carry out these exchanges long enough, they can establish correlations, which they can communicate to each other, between times of transmission and times of reception. For signals sent from M. Sanserif to Herr Muth it might be that $\mathbf{t} = \theta(T)$, while for signals sent from Herr Muth to M. Sanserif it might be that $T = \phi(\mathbf{t})$. There is no guarantee that these will be the same. But they can be made to be. That is, we can find two functions, ψ and h,such that

$$\psi^{-1}\left(h\left(h\left(\psi\left(x\right)\right)\right)\right) = \phi\left(\theta\left(x\right)\right) \qquad (4.4.1)$$

or, schematically to save brackets,

$$\psi^{-1}h^2\psi = \phi\theta \qquad (4.4.2)$$

or, equivalently

$$h^2 = \psi\phi\theta\psi^{-1}. \qquad (4.4.3)$$

Given the correlations θ and ϕ that they have established by their interchange of signals, M. Sanserif and Herr Muth can work out a procedure for recalibrating their clocks so as to have the same function connecting dates of transmission either way. They start by making M. Sanserif's clock "Muthworthy". Where previously he read T, now he applies the regraduating function ψ, and calculates $\psi(T)$, which works out at, say, W. Once M. Sanserif's clock is Muthworthy, it can be used to regraduate Herr Muth's clock by means of the rule that if a signal is sent from M. Sanserif at W, its arrival with Herr Muth is to be dated $h(W)$, which works out at, say, $\mathbf{w}$. If Herr Muth were thereupon to send it back, it would arrive at, say, $\mathbf{w}'$. What will be the relation of W' to $\mathbf{w}$? Before recalibration, M. Sanserif would have reckoned the date of dispatch to be not W but T, where $T = \psi^{-1}W$. Its arrival with Herr Muth would have been reckoned to be $\theta(T)$ before recalibration, and if immediately sent back, the return signal would arrive at $\phi(\theta(T))$ by the old reckoning. After M. Sanserif's recalibration, this needs to be regraduated by the function ψ; at the same time, instead of T, he has $\psi^{-1}(T)$. So

$$\begin{aligned} W' &= \psi\left(\phi\left(\psi^{-1}\left(W\right)\right)\right) \\ &= h\left(h\left(W\right)\right) \qquad (4.4.4) \\ &= h(\mathbf{w}). \end{aligned}$$

So, after recalibration, the relation between the dates of reception and transmission of signals from Herr Muth to M. Sanserif is the same as that between signals from M. Sanserif to Herr Muth, which is what we wanted to secure.

It might seem that we had destroyed symmetry in recalibrating Herr Muth's clocks by reference to M. Sanserif's time-signals. But this is not really so. For M. Sanserif's time-reckoning had first been groomed for the purpose, in that the ψ he used to recalibrate his own clock was itself determined by θ and ϕ, which in turn depended on Herr Muth's own clock readings. But if that were not enough to assuage Herr Muth's temporal chauvinism, we could equally well consider not $\phi(\theta(x))$ but $\theta(\phi(x))$ and find, together with a suitable regraduating function c, its functional square root, k, such that

$$\chi^{-1}\left(g\left(g\left(\chi\left(x\right)\right)\right)\right) = \theta\left(\phi\left(x\right)\right) \tag{4.4.5}$$

i.e.

$$\chi^{-1}k^2\chi = \theta\phi, \tag{4.4.6}$$

or, equivalently,

$$k^2 = \chi\theta\phi\chi^{-1}. \tag{4.4.7}$$

Either of these procedures would be enough to secure what we need, namely that the "signal function" shall be the same in both directions.

Once the Radio Rule is established as a symmetrical signal function $h(x)$, the rest of the argument goes through as in Section 4.2, except that instead of having a constant k and its reciprocal $1/k$, we have a function $h(x)$, and its inverse $h^{-1}(x)$. *M. Sanserif pense qu'il s'envoie un message à temps T_1, temps français, le message arrivera en Allemagne a temps t_2, temps allemand, où $t_2 = h(T_1)$. De même si Herr Muth respond à t_2, temps allemand, son message à lui sera de retour à temps T_3, temps français, où $T_3 = h(t_2)$. Ainsi $T_3 = h(h(T_1))$.*

But, granted uniform velocity, $L_2 = VT_2$, and

$$T_3 = \frac{1 + V/c}{1 - V/c}T_1. \tag{4.4.8}$$

Hence

$$h(h(T_1)) = \frac{1 + V/c}{1 - V/c}T_1, \tag{4.4.9}$$

and if this is to be generally true, $h(x)$ must be the function

$$\sqrt{\frac{1 + V/c}{1 - V/c}}x. \qquad (4.4.10)$$

By exactly the same reasoning $h(x)$ must also be

$$\sqrt{\frac{1 + \mathbf{v}/c}{1 - \mathbf{v}/c}}x, \qquad (4.4.11)$$

which, since $V/C = \mathbf{v}/\mathbf{c} = v/c$, we can write simply

$$\sqrt{\frac{1 + v/c}{1 - v/c}}x. \qquad (4.4.12)$$

After that, the argument goes through exactly as before.

We thus have derived the Lorentz transformation not from specific *ad hoc*, but from very general, assumptions which it would be hard to reject. Although there are still some assumptions implicit in the functional analysis which we have not fully articulated or justified—that the functions are continuous, differentiable, strictly monotonic, and therefore one-one with well-defined inverses—these cannot be seriously questioned without abandoning all our normal beliefs about continuity and causality. Both the Radar and the Radio Rule seem inescapable. The Radar Rule accepts that observers cannot at any one time be in any other place than where they are, and enables them to locate observed objects in their spatiotemporal conceptual scheme, subject to there being:

(i)   a finite maximum velocity of propagation of causal influence;

(ii) date-indifference: there is no preferred absolute dating system;

(iii) unit-tolerance: a change in the unit of time should make a proportional change in the measure of distance;

(iv) direction-indifference: outward and return journeys take the same time for the same distance.

The Radio Rule expresses a principle of equivalence—that two frames of reference moving with uniform velocity with respect to each other are equally "good", and assumes that the way M. Sanserif causes Herr Muth to receive a message is essentially the same as the way Herr Muth causes M. Sanserif to receive a message. In this respect M. Sanserif and Herr Muth each regard the other as observers, agents who can communicate, each being epistemologically as good as the other. If they regard each other not only as observers but also as objects of observation, they need to combine the Radio Rule with the Radar Rule, and so need to adopt the Lorentz transformation as the means of translating the spatiotemporal references of the one into the spatiotemporal references of the other.

§4.5 Critique

It is very surprising that conclusions as substantial as the Lorentz transformation should emerge from such general considerations as those governing the coherence of two different communicators schemes of reference. The reader will be suspicious, and ought to be. It seems too much like a conjuring trick to produce a rather empirical rabbit out of *a priori* air. Many philosophers hold that it is impossible to have valid *a priori* arguments for "synthetic" (*i.e.* substantial empirical) propositions. We need to look at the derivation with critical eyes.

Part of the difficulty is due to a confusion. Kant said that *a priori* truths were necessary, and many modern thinkers have supposed that there was only one sort of necessity, logical necessity which is defined in terms of inconsistency, as that which it would be inconsistent to deny. But there are many different sorts of necessity—legal necessity, social necessity, moral necessity, to mention only a few from outside science—and the distinction between *a priori* and *a posteriori* propositions, whatever it may be, is not the same as that between analytic and synthetic, or logically necessary and logically contingent, propositions. We leave until Chapter 10 discussion of non-deductive reasoning in physics, and here merely point out that while it would be clearly out of the question to deduce substantial empirical propositions from purely analytic—*i.e.* tautological—ones, it is not obviously impossible to bring forward other *a priori* arguments in support of them, and to derive specific mathematical results from general, though not tautological, assumptions.

In fact, the communication argument does not conjure empirical conclusions from purely non-empirical premises. It makes a number of substantial, though general, assumptions. Epistemological equality is thoroughly elitist: only a select class of observers—those moving with uniform velocity with respect to one another—are given parity of epistemological esteem. The principle of causality is invoked not only to establish that the time of reception of a radio message is a function of the time of its transmission, but that it is a strictly monotonic, and hence one-one, function. It is assumed that distance is a constant function of time, both in the Radar Rule and in the equation $l_2 = vt_2$. Moreover only one dimension of space is assumed, which as we saw in the previous chapter is not always an innocuous assumption.[12] The regraduation of the time-scales of the two observers, although it always can be done, since time is one-dimensional, is not a trivial matter, and might carry with it heavy costs in awkwardnesses elsewhere. The derivation is from empirical, but general, assumptions to empirical, but much more detailed, conclusions. It does not distil empirical results out of thin air, but shows how general constraints can entail precise solutions. The assumptions, as well as the general possibility of the argument, may also be criticized: Angel questions whether it can be taken for granted that there is a finite maximum velocity for the propagation of causal influence.[13] He points out that Action at a Distance, however awkward conceptually, is not a contradiction in terms, and that Newtonian mechanics, which admits infinite velocities, is not inconsistent. Hence, he argues, the communication argument must fail, or it would have ruled out Newtonian mechanics *a priori*.

But this is not so. Newtonian mechanics is a special case of the Special Theory, in which c is taken to be infinite, and the Lorentz transformation becomes the Galilean transformation. If we assume an infinite velocity, then the communication argument yields the Galilean transformation and a simple Newtonian view of time flowing uniformly and equably, independent of anything else.[14] Indeed, it is an intuitive communication argument adopting this assumption which makes the Newtonian view seem so compelling.

In any case, the consistency of the opposing view does not impugn the cogency of the argument. Action at a Distance may be

---

[12] In §3.2, pp.90-91, §3.3,p.94-96; see also below, §5.3, p.164, §5.6.

[13] R.B.Angel, *Relativity: The Theory and its Philosophy*, Oxford, 1980, pp.110–115.

[14] See above, §2.6, pp.46-47, and below, §8.6.

non-self-contradictory, but still is awkward,[15] and we may be led by arguments of continuity to a universal speed, as we were in Chapter 1; and Newtonian mechanics, although consistent, may none the less lead us on naturally to the Special Theory. Once rationality is not construed in terms of deductive logic alone, we can admit arguments in favour of a finite universal speed without having to make out that the contrary supposition is self-contradictory.

Angel also criticizes the assumption that the means of signalling actually available to us in radar and radio communication is the best there could be. Is it not rather complacent to suppose that we have the very best there could possibly be? Could there not be some faster speed we could use for sending messages? Indeed there could, as we have discussed in Chapter 3.[16] And the identification of electromagnetic radiation with the fastest possible means of signalling is always vulnerable to future falsification. We could be wrong, but are not wrong-headed, in making the identification. It is an empirical proposition, though a very reasonable one; and if it should prove to be wrong, and some other faster method of signalling be available, some non-instantaneous telepathic communication say, the argument would still go through, though with a different value of c, so long as observers could not only communicate with, but also observe, one another.

It could be the case, as Angel urges, that there was no finite maximum velocity or universal speed, and that we muddled along with the means of communication that were available to us as best we could. Then, certainly, there would be no communication argument. The communication argument can work only if we have a moderately sharp concept of communication and communicators, and such a concept may well go beyond our everyday experience, and be to some extent idealised. If it is wrong, or wrongly idealised, it can be criticized; but not for being idealised as such. Angel objects, finally, that there is an element of theodicy in the admission that it is a fortunate fact that the Lorentz transformation established by the communication argument coheres with the one under which electromagnetic phenomena are covariant, and the suggestion that ours is the most rational of all possible worlds. But any

---

[15] See below, §10.5; see also J.R. Lucas, *Space, Time and Causality*, Oxford, 1985, chs. 3,9, pp.40–42, 176–177.

[16] § 3.7.

scientist who is neither an extreme rationalist nor a radical empiricist is committed to some such admission. Since he is not an extreme rationalist, he acknowledges that experimental evidence can falsify his theories, and that it is logically possible for them to be false. But unlike the radical empiricist, he believes that his theories do more than just describe empirical observation, and aim to explain and reveal the inner harmonies of nature. In so far as he is successful, he is discovering the world to be rational: but it is not a foregone conclusion that it should be, in that it is not logically necessary, but only a fact, though in the view of any scientist who is seeking to discover the rationale of the world, a very fortunate one.

Nevertheless it is only a fact, and could have been different. The communication argument is not a deductive one, nor are its conclusions tautological. It is instructive, therefore to consider ways in which the communication argument could fail to deliver the results the physicist needs. It could be that the regraduation was at fault. Although any function onto the positive real numbers could do duty as a time-scale, most are unsuitable because they do not assign equal intervals to periodic processes. It could conceivably turn out that when we regraduated to secure a uniform Radio Rule, we produced a time-scale that was discordant with all natural processes. It would not be much good deriving a Lorentz transformation in which the frequency of sodium light was different in different frames of reference. It would, if the Lorentz transformation were derived by the communication argument, be a contingent fact that electromagnetic phenomena were the same in both frames of reference. It could also be that the Equivalence Principle is false, and the class of frames of reference which are moving uniformly with respect to one another are not equivalent.

There might be no parity of epistemological esteem between observers, and only one preferred frame of reference; or it might be that there were equivalent frames of reference, but that they were not related by the function $l_2 = vt_2$ but by some other function, possibly a lot more complicated. Or... , or... , or—there are many ways the derivation could be faulted. Yet each such objection is itself open to objection. Parity of esteem between different observers is, almost, a constitutive condition for the cooperative enterprise of scientific research. If we have any idea of distance, it is likely to affect how long messages take to reach their destination, and its time-derivative is likely to have some bearing on the function, which we have called h, that relates times of arrival to those of departure.

That is, h is likely to be, in the absence of the simplifying assumptions we made, a function of l and of v. Moreover, v should enter into h antisymmetrically. If M. Sanserif regards Herr Muth as moving away to the East with velocity v, Herr Muth should regard M. Sanserif as moving away with a velocity of equal magnitude but in the opposite direction, *i.e.* to the West; that is to say at velocity $\mathbf{v} = -V$. Hence h^{-1} should differ from h in that the terms containing v should change sign, and granted that h can be expanded as an analytic function, the terms in v must be linear, cubic, quintic, or some other odd power of v. Simplicity strongly suggests linearity. Similarly with each of the other objections, we cannot rule it out altogether, but can reckon that more argument is needed to establish the objection than to justify ignoring it.

The communication argument gives us a derivation which is not a water-tight mathematical proof, but a schema of argument which has many holes in it that a critic can, and should, pick; but having picked them, should also see how they may be blocked. No inconsistency results from refusing to accept its conclusion, but the conceptual price of doing so, either in complication, or in distortion of our basic concepts of time, space and causality, renders it an unattractive option.

§4.6 Metaphysical Suggestions

The communication argument is only one of many ways of deriving the Lorentz transformation, but it is particularly interesting philosophically, not only in deriving substantial conclusions from *a priori* premisses, but because it captures some presuppositions of contemporary thought, and articulates them into scientific form.

The Parity of Esteem enshrined in the Radio Rule is a principle implicit in our recognition of other minds and conjugation over persons. In construing your winces and drawn features as signs of your being in pain, I am "projecting" myself into your position and imagining how I would have to be feeling to evince behaviour like yours. Equally in addressing you in the second person or talking about you in the third person by means of a personal pronoun, I am ascribing to you the capacity of being an entity with a mind of his own, capable of feeling and rational decision, and entitled to use the first person to refer to your self: you are someone who can use the word 'I' of yourself. Although you are distinct from me and your point of view not mine, we are all of like passions with one another, and our points of view similar in important respects. The exact points of resemblance vary, but the principle of our all sharing a common humanity is important in literature and ethics, and that of our all being communicators in logic and metaphysics. It is the latter principle the Radio Rule takes up. The principle does not claim that we are all altogether the same. Rather, it claims that each of us has a first-personal perspective, which others can enter into and recognise the validity of, even though it is not their own. I cannot feel a pin in your leg, but I can recognise that a pin in your leg hurts you in much the same way as one in mine hurts me. Equally M. Sanserif does not have the same experience of sending and receiving messages as Herr Muth, but can recognise that he will use his messages to assign dates and distances to events in the same way as he, M. Sanserif, does, and that therefore any message function correlating the sending a message by one of them and its being received by the other should be the same for both, in something of the same way as both dislike pain. If M. Sanserif believes that the sending of a message by him a date T_1 results in its being received by Herr Muth at date t_2, and that the sending of a message by Herr Muth at date t_2 results in its being received by him at date T_3, then he will be under rational pressure to have t_2 being the same function, h, of T_1 as T_3 is of t_2. This rational pressure will be the same sort of pressure as leads us to reckon that you should seek to avoid pain, and try to maintain your freedom and enhance your ability to make further choices as I should.

Many philosophers have found it puzzling that we can project ourselves into other people, and can start useful thoughts with the necessarily false hypothesis 'If I were you...' Our ability to move out from our own particular personal point of view is less puzzling when we compare it with our ability to consider things from temporal standpoints other than our actual one. I can not only use the present tense to report things as they are now, but as well as the future and past tenses, the pluperfect and future perfect tenses, among others, which view events not from the present standpoint but some past or future one.[17] Conjugation over persons is not much more puzzling than conjugation over tenses.

The communication argument has a Leibnizian and humanist flavour in contrast to Newton's unitarian theism. For Newton, not only are all points alike in the sight of God, but only the God's-eye view is valid. God is omnipresent, and this is taken, as in Psalm 139, to make any other point of view merely partial and incomplete, and hence invalid. In the communication argument, however, the point of view of each different observer, though partial, representing just one particular frame of reference and not any others, is perfectly valid, and cannot be subsumed or replaced by any more general one. If we want to generalise, we do so not by moving away from a particular point of view, but by enabling an observer to take into account the fact that there are other points of view, and to project himself into them and accommodate his own thinking to them and *vice versa*. Instead of there being one uniform space which is the sensorium of God, with God being omnipresent throughout it, we think of there being a collocation of monads, in Leibniz' metaphysics, or a communion of souls in McTaggart's, which have established a certain harmony between them. Since the Monads are not windowless, and can observe as well as empathize, the constraint on possible harmonizations is stringent, so stringent, granted other rational principles, that only the Lorentz transformation will allow them all to sing in tune.

Newton's theism was simple, almost *simpliste*, so much so that he feared he would be thought heretical: and Leibniz was far from being an atheist. If we were to try and read anything off from the world-view enshrined in the communication argument, it would be one that laid great stress on the fundamental significance of communicators sharing a common rationality, but argued that if we want to escape, as we all must, from our egocentric predicament, we should do so not by aspiring to an impersonal world-view that denies or downgrades the validity of the individual's point of view,

---

[17] For an illuminating account, see Hans Reichenbach, *Elements of Symbolic Logic*, New York, 1947, §51, pp.287-298.

but an omnipersonal one that enables each to validate his own in also recognising the validity of those of others.[18]

---

[18] For an attempt to work out this line of argument in other areas of philosophy, see Thomas Nagel, *The View from Nowhere*, Oxford, 1986. On p.6 he reckons that his approach should have implications for the philosophy of space and time.

Chapter 5
Derivations of the Lorentz Transformation

§5.1 Synopsis

There are very many different ways of deriving the Lorentz transformation. We have given four already, and there are many others, as shown in Figure 5.1.1. The fact that there are so many, starting from from various, very general and sometimes rather vague, assumptions and yielding specific results, raises important philosophical issues. It gives a stereoscopic view of the Lorentz transformation, and leads us to regard it as the solid centre-piece of the Special Theory. It shows us something about the nature of physical and philosophical reasoning. And it reveals something of the interconnectedness of our conceptual structure, in that different assumptions lead to the same conclusion, and if any particular assumption is called in question, it can be defended in a variety of ways on the strength of others which have not been impugned.

Einstein was led[1] to the Lorentz transformation as the one that leaves Maxwell's equations unaltered in form in a transformation from one frame of reference to another (see Figure 5.1.2). For the physicist, with his feet firmly on the ground, this, along with the observed consequences of the dilation of time,[2] remains the most natural and compelling approach (see Figure 5.1.3). But though the covariance of Maxwell's equations under the Lorentz transformation remains a cogent reason for regarding it as physically important, it does not fully explain why it is important, or give an overall conspectus of the significance of the the Special Theory. Indeed, it is possible, as we shall show in Chapter 6, to reverse the order of argument, and derive Maxwell's equations from the Lorentz transformation, itself established on other grounds (See Figure 5.1.4). This is not to belittle the importance of empirical evidence or the significance of Maxwell's equations in their own

---

[1] See A.d'Abro, *The Evolution of Scientific thought from Newton to Einstein*, 2nd ed. New York, 1950, pp.135-136 .

[2] See above, §2.10, p.68.

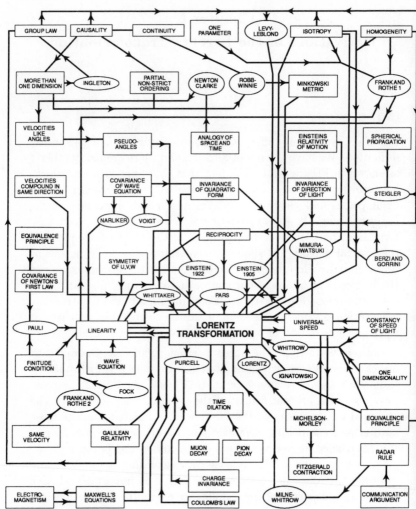

Figure 5.1.1   Ways to the Lorentz transformation. This diagram shows schematically the interrelationships of some of the arguments in this and other chapters. It is intended only as a broad overview. The concepts invoked are in boxes, but note that they are defined and used differently in different approaches. Because only two dimensions are available on the diagram, the Equivalence Principle is given a multiple location. The names in balloons indicate the authors of the different arguments. Some arguments in the text have been omitted to avoid over-complication.

right, but in order to show that they are not just brute empirical facts: they are, rather, facts that are not only empirically true, but also rationally coherent with other, very general and pervasive, features of space, time, spacetime and causality.

Figure 5.1.2 The young Einstein realised that the Lorentz transformation is that under which Maxwell's equations are covariant.

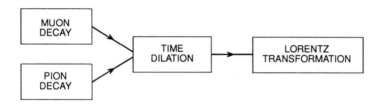

Figure 5.1.3 Empirical Approaches to the Lorentz transformation.

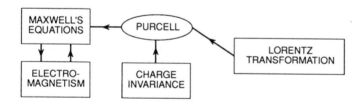

Figure 5.1.4 In Chapter 6 we shall follow Purcell's derivation of Maxwell's equation from the Lorentz transformation.

We have already given four derivations of the Lorentz transformation in the first four chapters of this book. These are shown schematically in Figures 5.1.5-5.1.8, and will be discussed further in §5.7. In the next sections we shall report a number of other derivations that have been offered by different authors. Some are difficult to follow. Often their authors themselves are not quite clear what exactly their assumptions are. It is only under critical probing that the precise content of a concept such as "homogeneity" becomes clear. Other basic concepts often appealed to are the universality or the constancy of the speed of light, the invariance of some quadratic form, various versions of the Principle of Equivalence,[3] often expressed in terms of group theory, certain other principles of symmetry, in particular that of linearity,and that of reciprocity, and finally, continuity.

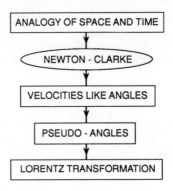

Figure 5.1.5 Integrationist approaches assume the fundamental importance of some quadratic form.

These concepts are not understood by every author in exactly the same way: we abbreviate them, and put them in boxes to emphasize this. Nevertheless, they provide an outline scheme for making sense of the welter of different derivations, which in this chapter we shall survey in order to show the many ways in which the Lorentz transformation is linked to other fundamental assumptions we make in thinking about the natural world.

---

[3] In this chapter we shall keep to the names used by the authors being considered. Often the principle they invoke—the "Principle of Relativity", or the "Galilean Principle of Relativity", or the "Galilean Relativity Principle"—is more restricted than in our use.

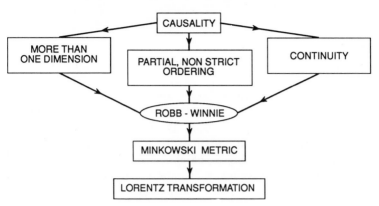

Figure 5.1.6 The Absolute Approach of Chapter 3.

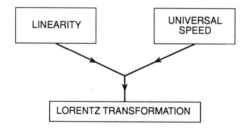

Figure 5.1.7 In Chapter 2 we derived the Lorentz transformation from Linearity.

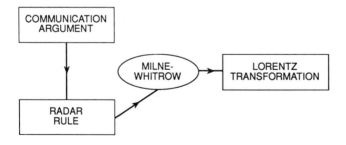

Figure 5.1.8 The Communication Argument of Chapter 4.

§5.2 Linearity

The very simple derivation given in Chapter 2[4] assumes that the transformation must be linear, *i.e.* that the equations are of the form $x' = Ax + Bt$; and $t' = Cx + Dt$, involving only the first powers of the variables x and t, and not any higher powers. In the same way, Whittaker argues that the function connecting compounded velocities must be linear.[5] On a graph, a linear function is represented by a straight line, in contrast to quadratics, which are represented by circles, ellipses, parabolas, or hyperbolas, and other, more complicated, functions which are represented by other, more complicated, curves. Besides its obvious simplicity and continuity, a linear function has the property of being strictly monotonic, and therefore one-one, with a unique inverse. Quadratic and sinusoidal functions are not strictly monotonic, for sometimes if x_2 is greater than x_1, $f(x_2)$ is not greater than $f(x_1)$, so that these functions are not one-one and do not have unique inverses. With transformations from one frame of reference to another, there must be inverse transformations, and so any functions involved must be one-one, and hence strictly monotonic. Linear functions thus emerge as strong candidates, but they are not the only functions that are monotonic and possessed of unique inverses: Ax^3, or $\sinh(x)$ satisfy these conditions also. Such functions are, admittedly, awkward, and can be ruled out on other grounds. But the simple argument from simplicity is not watertight, and it is natural to ask whether the requirement of linearity can be defended in some other way or replaced by some other assumptions.

Einstein in his original 1905 paper[6] appealed to the homogeneity *(Homogenetätseigeschaften)* of space and time, from which in conjunction with the Principle of Relativity and the constancy of the speed of light he derived the Lorentz transformation (Figure 5.2.1); but it is difficult to spell out precisely what homogeneity means. We may understand it simply as some sort of symmetry

---

[4] §2.7.

[5] §1.2.

[6] A. Einstein, *Ann. der Phys.*, **17**, 1905, pp. 898ff., tr. by W.Perrett and G.B.Jeffrey, *The Principle of Relativity*, London, 1923, p.44 of Dover edition.

under displacement; sometimes it has been taken to mean symmetry under re-orientation—that there is no preferred direction—but this is better described as isotropy; sometimes it has been taken to mean symmetry under change of scale—that there is no natural unit of length or of time. It is helpful to confine the term to the first sense, but even so there is still some further ambiguity: a spacetime of constant, but non-zero, curvature would seem to be homogeneous on many counts—every point in it would have exactly the same geometrical properties as every other point—but a flat spacetime, whose curvature is everywhere zero, is even more homogeneous, and it is an argument in favour of Euclidean space and Minkowski spacetime that because they are flat, they are more featureless and impose fewer constraints on the possible placement of things and the possible course of events than any other geometry.

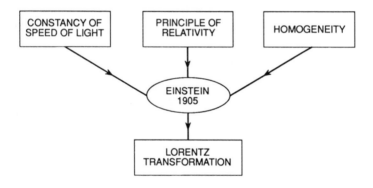

Figure 5.2.1 Einstein postulated the Principle of Relativity and the constancy of the speed of light, but needed Homogeneity to ensure that the transformation was linear.

Geometry, on this view, is sharply differentiated from physics. Geometry provides the framework of all possible characterizations of events, physics seeks to determine natural necessity. Geometry sets the scene, physics explains the actual course of events. Geometry therefore needs to be as little specific as it can be, in order not to constrain the realm of the possible, and to reserve all the explaining to physics.[7] And if we accordingly stipulate that spacetime be flat, we then have the Lorentz transformation straight away.[8] But this was not Einstein's view. Instead of differentiating between geometry and physics, he sought to identify them, and in his General Theory he put forward the programme of geometrodynamics, in which the physical explanation of gravity is entirely in terms of the geometrical structure of spacetime. It seems reasonable, therefore to understand Einstein's *Homogenetätseigeschaften* in the first, weaker sense. It would be enough to secure linearity that space and time be "origin-indifferent": it does not matter which particular point or which particular date we take as origin of our co-ordinate system, because all places and all times are alike, so far as the laws of physics are concerned.[9] If transformations between one co-ordinate system and another are to be origin-indifferent, they must be linear: and therefore, if we are to work with frames of reference at all, we must assume linearity, for only so will a change of origin in one co-ordinate system into another in the same frame of reference not alter the form of physical laws.[10] And hence also, if there is a transformation between one inertial frame of reference and another that makes no difference to the form of physical laws, it too must be linear, since only then would a change of origin in any co-ordinate system in one frame of reference always transform into merely a change of origin in the corresponding co-ordinate system in the other frame of reference. Linearity can thus be defended not only on the score of simplicity, but as already assumed in our use of the concept of a frame of reference, as an assumption which articulates a deep spatiotemporal symmetry built into the programme of physics as the search for those features of the world that are the same everywhere and everywhen.

Linearity can be defended in other ways too. It is reasonable to reckon that any transformation between frames of reference that is going to leave the form of physical laws unaltered must in particular preserve Newton's first law of motion that every material

[7] The choice is not always an open one. See Graham Nerlich, "What Can Geometry Explain?", *British Journal for the Philosophy of Science*, **30**, 1979, pp.69-83.

[8] Michael Friedman, *Foundations of Space-time Theories*, Princeton, 1983, p. 141.

[9] See above, §§2.4, 2.5, pp.38-43.

[10] See above, §4.3, p.135.

particle continues in a state of rest or uniform motion in a straight line: in which case straight lines must be transformed into straight lines (and, to preserve the distinctness of different straight lines, points at infinity into points at infinity). By restricting himself to transformations that transform straight lines and points at infinity into straight lines and points at infinity, Pauli was thus able to avoid having to assume linearity explicitly.[11]

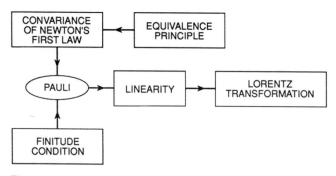

Figure 5.2.2   Pauli argues for Linearity from the restricted Equivalence Principle that requires only the Covariance of Newton's first law of motion.

In effect, he was concerning himself only with those transformations that resulted in *all* rectilinear motions in one frame of reference being transformed into rectilinear motions in another, and those transformations, unsurprisingly, are linear.

Linearity can be secured, not as reflecting a fact about the homogeneity of spacetime, but as a stipulation about the conditions of applicability of the Special Theory.[12] In addition to assuming that

---

[11] W.Pauli, *Theory of Relativity*, 1921, tr. G. Field, Pergamon, 1958, Part 1, §4, p.9. Y. Mimura and T. Iwatsuki, *Journal of Science*, Hiroshima University, **A, 1**, 1931, p.112, criticize him because "the assumption in his argument is too great, for whether all the rectilinear motions in one system would be the same in another system is quite beyond the scope of our observation". But this criticism would tell against any universal law of nature. If Newton can propound his first law of motion, Pauli can argue that it ought to hold in every inertial frame of reference, and that only transformations under which it is covariant are worthy of consideration.

[12] See below, §10.4.

the transformations relating to different pairs of observers must form a group, Pars needed "two conditions defining the particular problem to which the Lorentz transformation applies:

(1) Space and time are isotropic, and the equations of transformation are linear. We know from the General Theory of Relativity that space and time are not always isotropic, and that a general non-linear transformation is permissible when regard is paid to gravitational fields. The Lorentz transformation does not apply in the more general case; it is the solution of a special problem, and we cannot dispense with the above conditions which define that problem.

(2) Two observers agree as to the measure of their relative velocity. It is clear that we must introduce some convention in order to coordinate the two systems of measurement—there would be no significance in the existence of an invariant velocity for observers whose systems of measurement were entirely independent. The convention we choose is that they assign the same numerical measure to their relative velocity. It will be remarked that to establish the Lorentz transformation, *no further coordination of the measures is needed*."[13]

Pars proved that the composition of velocities is commutative, that is that $u + v = v + u$; he assumes that $u + (-u) = 0$, or equivalently that $-u + (-v) = -(u + v)$, an assumption fairly defensible on grounds of symmetry. Basically he argued from isotropy, linearity and reciprocity to the constancy of the speed of light and the Lorentz transformation, granted that all inertial frames of reference are equivalent, and the transformations relating them must form a group (See Figure 5.2.3).

Pars' assumptions were somewhat similar to those of Whittaker, whose derivation we made use of in Chapter 1.[14] Whittaker's assumptions (See Figure 5.2.4) were:

(1) If two velocities are in the same direction, then so is their resultant.[15]

[13] L.A.Pars, "On the Lorentz Transformation", *Phil. Mag.*, **42**, 1921, p.250.

[14] §1.2, pp.7-8.

[15] In this we are essentially ruling out the possibility of an Oersted-type observation. See below, §10.7.

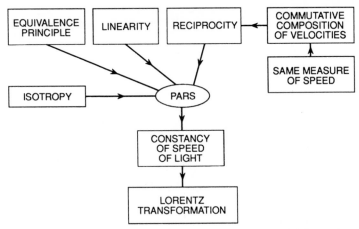

Figure 5.2.3 Pars proved the constancy of the speed of light, and hence the Lorentz transformation, from the Equivalence Principle, Linearity, Isotropy and agreement as to relative velocity.

(2) If u is the velocity of A with respect to B, then $-u$ is the velocity of B with respect to A.

(3) The function $g(u, v, w)$ which connects the velocities of A with respect to B, B with respect to C, and C with respect to A, is symmetric in u, v, and w.

(4) The function $g(u, v, w)$ which connects the velocities of A with respect to C, C with respect to B, and B with respect to A, is linear in u, v, and w.

And we can defend these assumptions, and in particular that of linearity, not as expressing facts about spacetime but as stipulations it is reasonable to make until the evidence forces us to revise them.

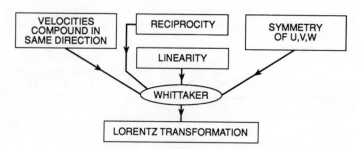

Figure 5.2.4 Whittaker used similar assumptions, but used them (§1.2, p.8) to argue for the general form of the combination rule, and thus for a universal speed that need not be infinite.

§5.3 Quadratic Forms

Linearity is a strong assumption, and although it can be defended, it is natural to ask whether it can be dispensed with, or itself derived from other assumptions. It is convenient first to consider derivations of the Lorentz transformation based primarily on the invariance of some "quadratic form".

The non-infinitesimal quadratic form,[16]

$$(x_1 - x_2)^2 + (y_1 - y_2)^2 + (z_1 - z_2)^2 - c^2(t_1 - t_2)^2, \qquad (5.3.1)$$

expresses the spacetime separation in Minkowski spacetime, and the invariance of

$$\sigma^2 = x^2 + y^2 + z^2 - c^2 t^2 \qquad (5.3.2)$$

is characteristic of Minkowski spacetime. The Lorentz transformation can be derived from the invariance of the spacetime separation, which expresses the view of Minkowski, that what is important is not space by itself and time by itself, but rather spacetime. But that is a view we did not actually come to until we were persuaded

---

[16] In this and the following sections, in order to follow more closely the authors we are citing, we reverse our previous convention of §1.5, p. 17, and give spacelike co-ordinates positive, and timelike co-ordinates negative co-efficients, and characterize the Lorentz signature as $(+ + + -)$ instead of $(+ - - -)$.

of the Special Theory on other grounds, and even if we are antecedently disposed to believe in spacetime, there is no reason why we should adopt the Lorentz signature $(+ + + -)$ or insert the factor c^2. To defend the particular way the separation is expressed we need to go further, and maintain something special about the speed of light; for instance, that it is the same in all inertial frames of reference. Whitrow[17] gives a simple derivation—an early version of the communication argument—of the Lorentz transformation in a one-dimensional world from the two assumptions:

(1) Einstein's postulate that for all sources and observers in uniform motion the speed of light be invariant, and

(2) Galileo's and Newton's postulate that a family of observers in uniform motion with respect to one another has no "privileged" members.

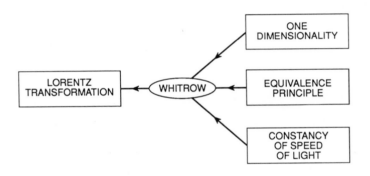

Figure 5.3.1 Whitrow derived the the Lorentz transformation from the Equivalence Principle and Constancy of Speed of light.

In his 1922 paper Einstein derived linearity from the invariance of $dx^2 + dy^2 + dz^2 - c^2 dt^2$ together with the "relativity of motion".[18] The invariance of $dx^2 + dy^2 + dz^2 - c^2 dt^2$, which is a quadratic form, is a slightly stronger assumption than that the speed of light should

[17] G.J.Whitrow,*Quarterly Journal of Mathematics* **4**, 1933, pp.161ff.

[18] A. Einstein, *Vier Vorlesungen über Relativitätstheorie*, 1922, p. 21. Einstein actually considered the invariance of $\Delta x_1{}^2 + \Delta x_2{}^2 + \Delta x_3{}^2 + \Delta x_4{}^2$. His approach is the four-dimensional one of §1.4.

be constant: the latter requires only that $dx^2+dy^2+dz^2-c^2dt^2 = 0$ along light rays, whereas the former holds of every infinitesimal separation. It was by having this slightly stronger assumption that Einstein was able to dispense with his appeal to homogeneity.

Figure 5.3.2 Einstein in his 1922 paper relied on the invariance of a quadratic form in *lieu* of Homogeneity to secure Linearity.

Narliker was able to strengthen the assumption about quadratic forms in another direction.[19] He considered the generalised wave equation

$$\frac{\partial^2 \phi}{\partial x_1^2} + \frac{\partial^2 \phi}{\partial x_2^2} + \frac{\partial^2 \phi}{\partial x_3^2} + \cdots + \frac{\partial^2 \phi}{\partial x_n^2} = 0. \qquad (5.3.3)$$

Although this equation asserts less than the general invariance of the left hand side, it is saying something about a function ϕ rather than simply laying down that a quantity $d\sigma^2$ should satisfy the invariance condition. In this way to say that it is "invariant" (that is, covariant) under a transformation is a stronger assertion, and from it Narliker was able to establish linearity for infinitesimal transformations, provided $n \neq 2$,[20] and hence the Lorentz transformation.

If we attach importance to equation 5.3.1, we might well reckon that only those transformations which left it essentially the same were of physical significance. But is it an important equation? It is the generalised form of the wave equation

$$\frac{\partial^2 \phi}{\partial x^2} + \frac{\partial^2 \phi}{\partial y^2} + \frac{\partial^2 \phi}{\partial z^2} - \frac{1}{c^2}\frac{\partial^2 \phi}{\partial t^2} = 0 \qquad (5.3.4)$$

---

[19] V.V. Narliker, "The Restriction to Linearity of the Lorentz Transformation", *Proc. Camb. Phil. Soc.*, **28**, 1932, pp. 460–462.

[20] See further below, §5.6.

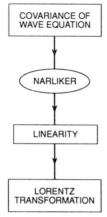

Figure 5.3.3 Narliker established Linearity from the Covariance of a
second-order differential equation

which it yields if we replace x_1 by x, x_2 by y, x_3 by z, and x_4 by ict.
Waves are in fact the way causal influence is propagated, and also
the means whereby signals are sent. They constitute the physical
substratum for the causal influence posited in Chapter 3 and the
communication of messages posited in Chapter 4. Narliker is assuming the physical importance of waves, whereas the approaches
of Chapters 3 and 4 assumed the philosophical importance of what
waves do. But does it have to be waves? Could causal influence
be propagated in some other way? Clearly it could. It might be
by impulse, the only way Locke could conceive of its happening;[21]
or it might be by instantaneous force, as in Newton's theory of
gravitation; or messages might be conveyed telepathically. These
cannot be ruled out as self-contradictory accounts of what might
happen. But they do rule out any possibility of further physical
explanation.[22] If we want a physically explicable account of the
propagation of causal influence, then the wave equation can be
plausibly offered as the simplest available. And then if we stipulate that the wave equation shall have the same form in every
inertial frame of reference, the Lorentz transformation will be the

[21] See below, §10.5, fn.19.

[22] See further below, §10.5.

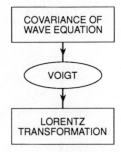

Figure 5.3.4   Voigt was the first to find the Lorentz transformation as that under which the wave equation $\Box^2 \phi = 0$ is covariant.

appropriate transformation to adopt.[23]

The constancy of the speed of light is, we saw, a slightly weaker condition than the invariance of $dx^2 + dy^2 + dz^2 - c^2 dt^2$; Mimura and Iwatsuki were able to derive the Lorentz transformation from this weaker assumption together with the principle of the relativity of motion and the invariancy of the direction of the ray of light which propagates parallel to the relative motion of the space co-ordinate systems.[24] In effect, by relying more heavily on quadratic forms, Mimura and Iwatsuki were able to establish Linearity on an extremely attenuated version of the Equivalence Principle in which only the paths of light rays (and not material particles as in Pauli's derivation[25]) have to be transformed into straight lines, and, indeed, only those light rays that are propagated in the direction of the relative motion of the space co-ordinate systems.

Instead of the covariance of the wave equation, Stiegler assumed that the shape of the wave front is the same in all inertial frames of reference.[26] He derived the constancy of the speed of light from

---

[23] This was, indeed, the very first route to the Lorentz transformation, noted by W. Voigt in 1887 (*Goett. Nachr.*, 1887, p.41), before Lorentz himself formulated them. See A. Pais, *Subtle is the Lord*, Oxford, 1982, p.121.

[24] Y. Mimura and T. Iwatsuki, "On the Linearity of the Lorentz Transformation", *Journal of Science*, Hiroshima University, **A**, **1**, 1931, pp. 111–116.

[25] §5.2, p.159 above.

[26] K.D. Stiegler, *Compt. Rend.*, **234**, 1952, pp. 1250ff.

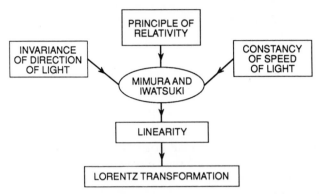

Figure 5.3.5 Mimura and Iwatsuki require the constancy of the speed of light and that the paths of light rays be straight in all frames of reference.

three assumptions, namely:

(1) space is homogeneous and isotropic,/

(2) the propagation of light is spherical, and

(3) the laws of physics are the same in all Galilean[27] frames.

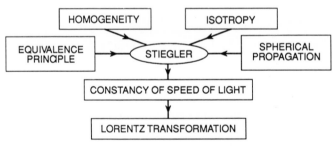

Figure 5.3.6 Stiegler needs only the invariant *shape*, instead of the invariant speed, of the wave front.

From these it is not surprising that he derives the constancy of the speed of light, since if the propagation of light is always spherical, it must have the same speed in all frames of reference, and if the laws of physics are to be the same in all Galilean frames of reference, the transformation will have to transform straight lines into straight lines, and thus be linear, and must depend only on v.

[27] See n.3 of §5.1, on p.154.

§5.4 Linearity and Quadratic Forms

The interplay between linearity and quadratic forms is illustrated by two papers by Frank and Rothe. In the first they were able to derive the constancy of the speed of light from linearity together with some assumption of isotropy.[28] They assume that:

(1) equations of transformation form a linear, homogeneous, one-parameter[29] group, and

(2) space and time are isotropic in so far as only the magnitude, and not the direction, of the relative velocity appears in the transformation equations.

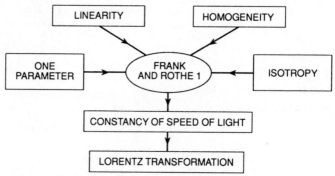

Figure 5.4.1 Frank and Rothe (1) argued from Linearity to Invariant speed of light, assuming various other sorts of sameness.

Both assumptions seem plausible, although Stephenson and Kilmister regard their assumptions of group properties as rather special-ised.[30] It is natural to think that only the velocity should determine the transformation between one frame of reference and another moving with uniform velocity with respect to it; [31] and that the

---

[28] P. Frank and H. Rothe, *Ann. der Phys.*, **34**, 1911, pp. 825–855; for a brief account, see G. Stephenson and C.W. Kilmister, *Special Relativity for Physicists*, London, 1958, pp. 19–21.

[29] See below, §5.5, p.176.

[30] G. Stephenson and C.W. Kilmister, *Special Relativity for Physicists*, London, 1958, p. 21; but see below, §5.5, p.175, and §5.6.

[31] See above, §2.7, §4.4.

general form of the transformation should be the same whatever the direction of the motion of the one frame of reference with respect to the other. Of course, it would not be plausible to make the time co-ordinate depend on the space co-ordinate unless the Special Theory had already been established on other grounds: but since the Galilean transformation can always be seen as a special case of the Lorentz transformation, with c being taken as infinite, that is not an objection to this derivation, which is simply more general than Newtonians would need.

In their second paper Frank and Rothe avoid having to assume linearity by deriving it from two further assumptions, the Galilean Principle of Relativity,[32] namely that the principle of inertia holds good in all inertial frames, and the assumption that if two particles are both assigned the same velocity in one inertial frame then in any other inertial frame the velocity assigned to the one must be the same as that assigned to the other.[33] If the principle of inertia holds in all inertial frames, straight lines must be transformed into straight lines, which is the assumption Pauli required in order to be able to derive the Lorentz transformations without explicit appeal to linearity. Linearity is being defended not by some appeal to date-indifference, but in order to ensure that Newton's First Law of Motion shall hold in the Special Theory as well as in ordinary Newtonian mechanics. The other principle, that if two particles are both assigned the same velocity in one inertial frame then in any other inertial frame the velocity assigned to the one must be the same as that assigned to the other, while not in the least implausible, is not obviously compelling if taken as an assertion of fact, but can better be seen as a stipulation analogous to the recalibration procedure of the communication argument.[34]

The first of these derivations, together with that of Narliker in the preceding section, shows that in the presence of certain other plausible assumptions, linearity is inter-derivable with the constancy of the speed of light, and the second that linearity can be

---

[32] See n.3 of §5.1 on p.154.

[33] P. Frank and H. Rothe, "Zur Herleitung der Lorentz Transformation", *Phys. Zeitschr.* **13**, 1912, pp. 750–753; summarised by R. Torretti, *Relativity and Geometry*, Oxford, 1983, p. 298, n. 3.

[34] §4.4

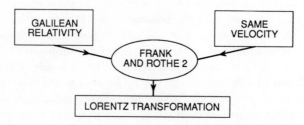

Figure 5.4.2 Frank and Rothe (2) no longer need to assume Linearity, which they derive from the Galilean Principle of Relativity together with the requirement that velocities be all transformed the same way.

defended on general Newtonian grounds: if the Special Theory is to be a refinement, rather than a refutation, of Newtonian mechanics, then the Lorentz transformation is the one under which it will have to be covariant.

The Russian physicist, Vladimir Fock, has examined the role of the two assumptions more rigorously. He formulates first a condition from Newtonian mechanics, that

$$
\begin{aligned}
x &= x_0 + v_x(t - t_0) \\
y &= y_0 + v_y(t - t_0) \\
z &= z_0 + v_z(t - t_0)
\end{aligned}
\tag{5.4.1}
$$

should transform into

$$
\begin{aligned}
x' &= x'_0 + v'_x(t' - t'_0) \\
y' &= y'_0 + v'_y(t' - t'_0) \\
z' &= z'_0 + v'_z(t' - t'_0)
\end{aligned}
\tag{5.4.2}
$$

and, second, an electromagnetic condition, that from

$$
\frac{1}{c^2}\left(\frac{\partial \omega}{\partial t}\right)^2 - \left[\left(\frac{\partial \omega}{\partial x}\right)^2 + \left(\frac{\partial \omega}{\partial y}\right)^2 + \left(\frac{\partial \omega}{\partial z}\right)^2\right] = 0
\tag{5.4.3}
$$

in the unprimed frame, there should follow in the primed frame

$$
\frac{1}{c^2}\left(\frac{\partial \omega}{\partial t'}\right)^2 - \left[\left(\frac{\partial \omega}{\partial x'}\right)^2 + \left(\frac{\partial \omega}{\partial y'}\right)^2 + \left(\frac{\partial \omega}{\partial z'}\right)^2\right] = 0
\tag{5.4.4}
$$

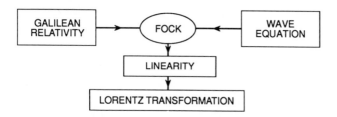

Figure 5.4.3   Fock derived Linearity from the two conditions that the Equivalence Principle should hold both for Newton's first law (Equation 5.4.1) and for the wave equation (Equation 5.4.3).

Granted these two conditions, the most general functions of the form

$$
\begin{aligned}
x' &= f_1(x, y, z, t) \\
y' &= f_2(x, y, z, t) \\
z' &= f_3(x, y, z, t) \\
t' &= f_4(x, y, z, t)
\end{aligned}
\tag{5.4.5}
$$

must be such that

$$
\frac{\partial^2 f_i}{\partial x_k \partial x_l} = 0,
\tag{5.4.6}
$$

which is the differential expression of the transformation being linear.[35]

Fock goes on to consider the consequences of dispensing with either of these conditions.[36] If we have only the first condition, the transformation could be one involving linear fractions, all with the same denominator, instead of a simple linear transformation. If we have only the second, the transformation could be not just a Lorentz transformation alone, but a Lorentz transformation together with a "Möbius transformation". We can rule out the latter, unwelcome, possibility by requiring that finite values of the initial co-ordinates should lead to finite values of the transformed ones. This is the additional stipulation that Pauli needed to make in

---

[35] V.A. Fock, *Space, Time and Gravitation*, tr. N.Kemmer, Oxford, 1959, § 8, pp.13–16; 2nd ed., 1964, § 8, pp.20–24.

[36] V.A. Fock, *Space, Time and Gravitation*, tr. N.Kemmer, Oxford, 1959, pp.377–384; 2nd ed., 1964, Appendix A, pp.403–409.

§ 5.2. Thus although the two conditions are conjointly sufficient, they are neither of them necessary for the derivation. Either in the absence of the other admits more than we want to admit, but this more can be winnowed down by further conditions ruling out the other unwelcome possibilities that do not make sense physically.

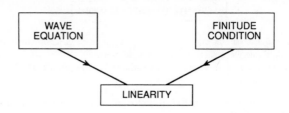

Figure 5.4.4   Fock can almost dispense with the condition that the Equivalence Principle hold for Newton's first law, but does then need a finitude condition.

§5.5 In Place of Linearity and Quadratic Forms

It is natural to press the enquiry further, and see if we can derive the Lorentz transformation without having to assume either linearity or the constancy of the speed of light. Ignatowski and his successors have produced a family of derivations of the Lorentz transformation in a general form without assuming either.[37]

Ignatowski himself claimed to be able to base his derivation solely on the Galilean Relativity Principle, that all inertial frames of reference are equivalent, but a closer analysis of his argument shows that he needs also a "reciprocity principle", that if one inertial frame has a uniform velocity v with respect to another, then the

---

[37] W.A. von Ignatowski, "Einige allgemeine Bermerkungen zum Relativitätsprinzip", *Verh. Deutsch. Phys. Ges.*, **12**, 1910, pp. 788–96; and W.A. von Ignatowski, "Einige allgemeine Bermerkungen zum Relativitätsprinzip", *Phys. Zeitsch.*, **11**, 1910, pp. 972–6 and W.A. von Ignatowski, "Das Relativitätsprinzip", *Arch. f. Math. u. Phys.*, **3**, 1911, § 17 pp. 1–24, and 18 pp. 17–41. For a full account, see R. Torretti, §3.7, pp. 76–82. See also H. Bacry and J.M. Lévy-Leblond, "Possible Kinematics", *J. Math. Phys.*, **9**, 1968, pp. 1605–1614.

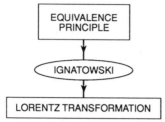

Figure 5.5.1 Ignatowski claimed the Equivalence Principle as sufficient for the derivation of the Lorentz transformation.

velocity of the second with respect to the first is $-v$. The principle of reciprocity seems highly acceptable; it can be justified by
an appeal to a principle of spatial isotropy along the direction of
uniform motion.[38] It was invoked explicitly as one of the symmetry conditions required for the derivation given in Chapter 2,[39]
and implicitly in the simple, though not in the full (being there a
consequence of the recalibration procedure), version of the communication argument.

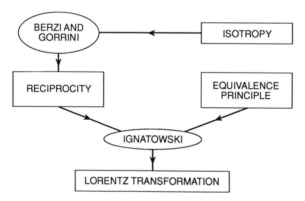

Figure 5.5.2   Ignatowski needs Reciprocity as well. It can be based
on Isotropy.

The principle of reciprocity is difficult to quarrel with, though it
presupposes a common standard of measurement that may not be

---

[38] V. Berzi and V. Gorrini, "Reciprocity Principle and the Lorentz Transformations", *J. Math. Phys.* **10**, 1969, pp. 1518–1524.

[39] Assumption (ii) of of §2.7, p.53.

available. The principle of spatial isotropy is invoked in setting the ε of Einstein's general Radar Rule equal to $\frac{1}{2}$,[40] and is even harder to quarrel with, since we do not have to posit it as a separate assumption, but can, on the Absolute Approach, derive it as a consequence of the topological structure of spacetime.[41]

Ignatowski's derivation has been closely scrutinised by Torretti, and refined by Bacry and Lévy-Leblond and by Berzi and Gorrini. A masterly presentation by Lévy-Leblond elucidates not only the structure of this derivation but of several others as well.[42]

Lévy-Leblond identifies four further assumptions that are needed for the derivation of the Lorentz transformation from the Equivalence Principle. The four assumptions are:[43]

(1) Isotropy of Space;

(2) Causality;

(3) the Group Law; and

(4) Homogeneity of Spacetime.

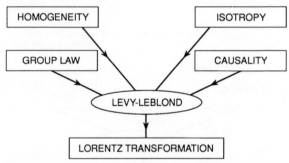

Figure 5.5.3   Lévy-Leblond derives the Lorentz transformation from Homogeneity and the Equivalence Principle, granted that the latter is expressed in a Group Law, together with a slight dependence on Isotropy and Causality.

---

[40] §2.8, pp.59-61, and §4.3, p.138.

[41] See above, §3.4.

[42] J.M. Lévy-Leblond, "One more derivation of the Lorentz Transformation", *American Journal of Physics*, **44(3)**, 1976, pp. 271–277.

[43] Lévy-Leblond's order has been changed for ease of exposition.

Of these assumptions, Isotropy of Space is needed only to establish the reciprocity principle; and Causality is needed only to rule out a bizarre possibility not otherwise excluded; it could be replaced by the condition that two velocities in the same direction cannot add up to a velocity in the opposite direction.[44]

Lévy-Leblond's third assumption is that the transformations form a group. This is what we should expect from the fact that we are dealing with an equivalence class of inertial frames. Group theory provides the fine structure of equivalence relations; and if the transformations did not form a group, they would not be operating within the equivalence class picked out by the Equivalence Principle. The group property itself is, thus, defensible on very general grounds; and though Stephenson and Kilmister claim that the further assumptions needed by Frank and Rothe are rather specialised, in the contexts of Lévy-Leblond's derivation they too can be defended on general grounds.

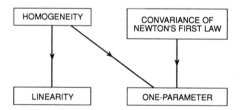

Figure 5.5.4   Lévy-Leblond argues that Homogeneity secures not only Linearity, but, with the covariance of Newton's first law, that the group of transformations should have only one parameter.

That the transformation is linear is secured by the fourth assumption, of Homogeneity. It also provides the basis for an argument that the set of transformations must have one and only one parameter (granted suitable alignment of axes: in the general case, of course, there will be three parameters, corresponding to the three components of the relative velocity), since if there were no parameter there would be no set of transformations at all, and if there were two or more, there would be a special set of co-ordinates under the transformation, contrary to the requirement of homogeneity that all co-ordinates are treated the same.

[44] A.R. Lee and T.M. Kalotas, "Lorentz transformations from the first postulate", *American Journal of Physics*, **43(5)**, 1975, pp. 435–436.

Moreover, as Lévy-Leblond also argues, unless there were one and only one parameter of the transformation, we should not have the covariance of Newton's first law, but should either be unable to transform one uniform velocity in one inertial frame of reference into a different uniform velocity in another inertial frame of reference or else be able to transform an acceleration in one inertial frame of reference into a different acceleration in another one. Thus there being just one parameter of the transformation is connected both with the homogeneity of spacetime and with the exact extent of the equivalence class picked out by the Equivalence Principle.

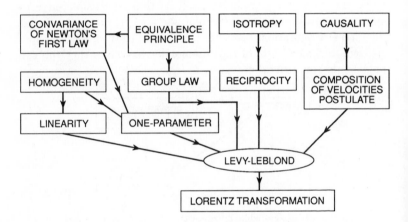

Figure 5.5.5   Lévy-Leblond's full argument. The requirement of causality is needed only to secure that two velocities in a given direction combine to form a velocity in the same direction.

§5.6 Dimensionality

Dimensionality is important in some of the derivations. Zeeman needed to have more than one spatial dimension, in order to ensure that parallel rays did not yield only the identity transformation,[45] and Narliker had to exclude the case where $n = 2$.[46] Ingleton made use of there being more than one spatial dimension to consider three inertial frames of reference, the third moving with respect to the second perpendicularly to the line of the motion of the second with respect to the first (Figure 5.6.1).[47]

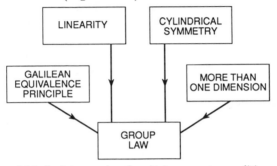

Figure 5.6.1 Ingleton needs only certain symmetry conditions, provided there is more than one dimension of space.

He was able to show that, granted linearity and cylindrical symmetry around the common axis of each pair of frames and allowing the possibility that the time co-ordinate may be dependent on the space co-ordinates[48] there is an absolute speed, independent of the magnitude and velocity of the second frame of reference with respect to the first; and hence that if we take this as constituting a measure of speed for all frames, we have the Lorentz transformation as a result. Only the Lorentz transformations form a group of the required type that will generate an equivalence class in more than one direction. What Ingleton's derivation achieves is to show

---

[45] §3.2, above, pp.90-91.

[46] This chapter, §5.3, p.164 above.

[47] A.W. Ingleton, *Nature*, **171**, 1953, pp. 618–619.

[48] See above, §5.4, pp.168-169.

that we do not have to assume that the linear transformations between different inertial frames of reference must form a group, as
Frank and Rothe do,[49] but that the group properties follow from
the condition of the Equivalence Principle holding in more than one
dimension. In that case the only linear transformation we can con-
sistently adopt that will work for a succession of transformations
in different directions is the Lorentz transformation.

§5.7 Review

Some common themes and family resemblances can be seen in these
different derivations. Some are integrationist in spirit. In Chapter
1 the Lorentz transformation was obtained as an immediate consequence of the programme of integrating space with time. If space
and time can be integrated into a single spacetime, there must be
some quasi-rotation between dimensions in spacetime analogous
to ordinary rotations in ordinary space, with velocities resembling
gradients and rapidities, or pseudo-angles, resembling angles. The
Lorentz transformation then emerges as the natural analogue to
ordinary rotations in Euclidean space. So, too, to assume that
$x^2 + y^2 + z^2 - c^2t^2$, or $\mathrm{d}x^2 + \mathrm{d}y^2 + \mathrm{d}z^2 - c^2\mathrm{d}t^2$, is the same in
all inertial frames is to assume that time is rather like, but not
exactly like, space; and a transformation, which is evidently not
a displacement but preserves magnitudes, is analogous to a rotation. Zeeman is likewise integrationist, in assuming the existence
of spacetime, with, as it happens, a Lorentz signature.

The approach is that of Minkowski rather than Einstein, geometrical rather than physical. It has great clarity and unifying
power, but these are purchased at the cost of introducing an entity
that many people find difficult. Space and time are difficult concepts on any count, and spacetime doubly so. "What is space?"
we ask; "what is time?" "Are they entities in their own right, or
are they something more insubstantial, perhaps just a system of
relations?" Space and time seem to be curiously tenuous and unreal. We are chary of assigning them a key role in any schema of
explanation, for fear they may prove unable to sustain the part
they are required to play.

Quadratic forms emerge not only as a four-dimensional version of
Pythagoras' theorem but in the form of the wave equation. Waves

[49] See above, §5.4, p.168.

are a plausible physical mechanism for the propagation of causal influence, if we accept continuity as a necessary condition of physical explicability. Narliker's derivation parallels the argument of Chapter 1, [50] basing an integrationist result on causal explicability. The constant spherical shape assumed by Stiegler and the constancy of the speed of light, or of its direction, are again "quadratic form" arguments based on the central role of the propagation of electromagnetic radiation in any physical theory of causality.

The approach of Chapter 3 is causal too, but on a large, not an infinitesimal, scale. Causal influenceability is an ordering, that is to say a transitive, asymmetric relation: and whereas with transitive, symmetric, that is to say equivalence, relations, there are many different respects in which physical systems may be regarded as the same, there is only one ordering relation that is a candidate for being of fundamental physical importance. The approach of Chapter 3, like that of Chapter 1, is geometrical, but it is much less expensive in the entities it has to postulate. It does not postulate the existence of spacetime, but only of spatiotemporal events which may, or may not, be able to influence one another. It has a certain purity of approach which makes up for its abstractness and remoteness from physical observation.

Communication is a form of cause, but the thrust of the communication argument is much more like that of Chapter 2 than that of Chapters 1 or 3. It is epistemological, concerned with parity of esteem, that is, some sort of sameness, between observers, each of whom is also by others observed. Because they are observers, all on the same footing, they are subject to the Radio Rule, which should be symmetric as between any two observers, and because they are observed, they are dated by other observers according to the Radar Rule. The Lorentz transformation emerges as the only way of harmonizing the Radio and the Radar Rule, each expressing some sort of symmetry between the two observers.

Symmetry is the underlying theme of Chapter 2 and the derivations based upon the Equivalence Principle, the reciprocity principle and linearity. It has the advantage of being much less metaphysical that the integrationist approach of Chapter 1, less abstract than the absolute approach, and less finicky than the communication argument. We do not posit the existence of some mysterious

[50] §1.1.

spacetime, but only ask under what conditions the same phenomena occur.

The Lorentz transformation emerges as that which preserves the important samenesses of electromagnetism. But sameness turns out to be a treacherous concept, and our understanding of the Special Theory is often needlessly confused because it is not made sufficiently clear with respect to what some phenomena are regarded as being, or not being, the same. Particular difficulty arises with simultaneity, meaning *occurring at the same time as*. In order to apply the relation *occurring at the same time as*, or *being in the same place as*, we have to specify some frame of reference which is essentially arbitrary, and having specified some arbitrary frame of reference, we are impelled to generalise again to a level above that arbitrariness. We want to talk not about this or that particular frame of reference, but about what is common to all. *Occurring at the same time as* and *being at the same place as* are not common to all, and are therefore not of great interest to physicists. Hence the emphasis on frames of reference that the approach of Chapter 2 engenders is likely to obfuscate the most important issues. It introduces a particularity that then has to be surmounted. What is really significant is spacetime separation, which is the same for all frames of reference, rather than duration or distance, which depend on the choice of a particular frame of reference; and it is the lightcone, rather than any particular frame of reference within it, that gives the key to the fundamental physics of any system.

In spite of these difficulties, considerations of symmetry are of great importance throughout physics, often because they are close to experimental observation. In the Special Theory the cardinal sameness is that between different frames of reference moving with uniform velocities with respect to one another. A second sameness is that of Reciprocity, which lays down that the velocity ascribed by one frame of reference to another should be the same, but in the opposite direction, as that ascribed by the other to the original frame of reference. A third sameness is linearity, deriving from the origin-indifference of the different co-ordinate systems within one frame of reference, and expressing the homogeneity of space and time. A fourth sameness, invoked by Lévy-Leblond to justify the Reciprocity Principle, and assumed by some others, is the isotropy of space.

We can usefully distinguish the roles these different samenesses play in derivations of the Lorentz transformation, if we ask three questions about the Lorentz transformation:

1. What is it of?
2. What is it between?
3. What has it got to achieve?

The answer to the first question is that, at least on a macroscopic scale, the Lorentz transformation is about points in space and time; it is a scheme for redescribing (or, more exactly re-referring to) spatiotemporal points. It must therefore be a one-one mapping onto with a unique inverse, or it would not be a redescription of space and time at all. And unless it were continuous, it would be completely inadequate. These considerations alone nearly suffice to give us linearity, and we can argue on very general grounds that have nothing to do with the Special Theory for the homogeneity of space and time.

The answer to the second question is that the Lorentz transformation is between inertial frames of reference, that is to say, frames of reference moving with uniform velocities with respect to one another. This is the equivalence class picked out by the Equivalence Principle. It might well have been a narrower or a wider equivalence class. There is no *a priori* reason why a uniform velocity should not make a difference to the form of physical laws, as a uniform acceleration does; and it is possible to conceive a physics in which angular velocities and uniform accelerations made no more difference than uniform velocities in fact do. That the Equivalence Principle should extend over frames of reference moving with uniform velocity with respect to one another and no further is characteristic of Newtonian mechanics and the Special Theory, and differentiates them from other physical theories.

The answer to the third question is that the Lorentz transformation has got to be such that the laws of nature are covariant under it: that is, if we transform from one frame of reference to another, the laws of nature must take the same form when transformed as they did originally. It is a moot question exactly which laws of nature should be Lorentz-covariant. Einstein thought that the Equivalence Principle applied to absolutely every truth, scientific and non-scientific alike, but as we have seen,[51] this is to

---

[51] See above, §3.9 p.118.

extrapolate too far. Even within physics there are difficulties in making quantum mechanics covariant under the Lorentz transformation, and the General Theory is avowedly not covariant under the Lorentz transformation. In fact there were only two physical theories, Newtonian mechanics and Electromagnetic theory, which the transformation between inertial frames of reference were required to preserve in the same form.

It may seem surprising that in some derivations of the Lorentz transformation it was enough that the principle of inertia should hold in all inertial frames of reference, and no reference was made to specifically electromagnetic phenomena. This is a further indication of how very close together Newtonian mechanics and the Special Theory are: the Lorentz transformation is just a generalisation of the Galilean transformation; or, to put it the other way, the Galilean transformation is a special case of the Lorentz transformation, in which c is infinite.

§5.8 Discussion

Faced with all these derivations of the Lorentz transformation, the working physicist is likely to feel somewhat exasperated. For him a theory is a tool to be used to help him understand the world, to make sense of the numbers that come from his measurements and to guide him in his plans for future experiments. If a theory is working well, he tends not to think much about its detailed justification, except when he is lecturing to students or writing a textbook. It is only when the theory starts to fail him that he will begin to scrutinise its foundations with anxious care. And so far the Special Theory has not failed him; it is triumphantly successful to whatever accuracy he can attain in his measurements, and this is often very high. So he is rather inclined to regard all these derivations of the Lorentz transformation as just an amusing game for philosophers or for physicists who regrettably cannot think of anything better to do. But there is more to it than that.

In the first place, we are being shown something about the nature of physical argument. Arguments in physics are, characteristically, not rigorous proofs in formal logic. Physicists, however, often feel that their arguments should be rigorous, and that they have failed in their job if they produce mere "hand-waving arguments". This is a mistake. There is no one ideal of scientific argument with everything else falling short of it, but rather a graduation of arguments, some more, some less rigorous, neither being better in any sense than the other, but different ones being suited to different occasions. A hand-waving argument may indicate the essential physical content of a derivation, where a full formal rigorous proof may succeed only in obscuring the shape of the wood by excessive emphasis on unimportant trees. On the other hand we may on occasion be concerned to specify more fully the background assumptions being taken for granted. It was perfectly reasonable for Einstein to appeal to the homogeneity of spacetime when he was arguing from the constancy of the speed of light, since it was the latter that differentiated the Special Theory from the theory of some luminiferous ether. But at a later stage we need to unpack the concept of homogeneity, and separate out the premiss that the curvature of spacetime is constant from the claim that it is always zero. The latter claim can be argued for within the context of electromagnetic theory (though not within that of gravitation theory): only in flat spacetime can size vary independently of shape; only in flat spacetime does Pythagoras' theorem or any simple analogue hold; the group-theoretical treatment of flat spacetime is simpler than that of curved, even constantly curved, spacetime. Einstein's assumption, although not explicit, can nevertheless be defended. Total explicitness is not called for, indeed is unattainable. Many modern mathematicians are almost unintelligible unless the reader has already mastered some less rigorous treatment, and can appreciate the point of labouring to prove rigorously what to any normal physicist would anyhow be obvious. There is no virtue, either pedagogic or academic, in obscuring the wood by distinguishing too many trees: it is enough that any bit of the wood, on closer examination, can be seen to be constituted of genuine trees. That is to say, all that is required is that assumptions can be made explicit and challenged and defended if need be. A holey derivation is acceptable so long as the holes are acknowledged, and can be plugged if required.[52]

In the second place, the very variety of the derivations throws light on the basis of physical theories. Physicists are constantly thinking up for themselves what they suppose to be new derivations of the Lorentz transformations, often "unaware they had long

---

[52] See, more fully, §§ 10.2, 10.3 below.

been available in well-known journals".[53] This is because they often come to the Special Theory from slightly different standpoints, taking different principles for granted, and regarding other principles as questionable. Since the Lorentz transformation occupies a crucial position, being the keystone of all calculations in the Special Theory, each physicist needs to accept it on his own terms. Hence, as he thinks through other derivations of the Lorentz transformation, he is likely to find them unsatisfactory, either in making assumptions that are not justified, or in obscuring the issue by proving points that do not need to be proved; and so he seeks a derivation which he can regard as satisfactory, because it starts from exactly the position he occupies, and leads to the result he needs to arrive at.

In the third place, the variety of the derivations shows how fundamental physical principles may be defended against the philosophical sceptic. In the orthodox empiricist philosophy, the only answer to the sceptic who queries the Lorentz transformation is to appeal to experiment. But that, though often effective, is unilluminating. It shows the sceptic to be wrong, but does not show him why he is wrong. But now that we have shown the Lorentz transformation to be rooted in our conceptual structure in a wide variety of ways, we can obtain a grip on the sceptic, and show him what else he would have to give up if he were seriously minded to deny the validity of the Lorentz transformation. Different sceptics have different doubts, but also different convictions. If a sceptic asks us to justify the Lorentz transformation, we can ask him what he is prepared to accept. If he accepts that time and space resemble each other in important ways and that it is a reasonable goal of scientific enquiry to develop an integrated theory of space and time, then the Lorentz transformation comes out as the natural analogue of rotation in purely space-like space. If he accepts that time and space are so homogeneous that it should not be possible to determine by experiment whether one is at rest or not, then Einstein's derivations can be appealed to. Whatever his view of physics he is prepared to hold, it is likely to furnish the assumptions needed to enable some derivation to proceed. The Lorentz transformations can be defended not as an opaque brute necessity, which just happens to be the case, but as rationally required by

[53] Roberto Torretti, *Relativity and Geometry*, Oxford, 1983, p. 76.

other views about space, time and causality, at least some of which the sceptic is likely to share.

The objection may be made that the sceptic need not share any assumptions at all, but may put all of them equally in question, after the manner of the Cartesian or Humean sceptic. Hume puts the point well:

> And if every attack, as it is commonly observed, and no defence ...... is successful, how complete must be his (*i.e.* the sceptic's) victory who remains always, with all mankind on the offensive, and has himself no fixed station or abiding city which he is ever, on any occasion, obliged to defend?[54]

But such a scepticism is, on Hume's own admission, indefensible. No alternative position is being maintained, and if no position can be maintained, then the sceptic is abandoning not just the pretence of actually possessing knowledge, but all hope of ever attaining to it. To adopt a policy which precludes the possibility of ever knowing anything is unappealing. Although any doctrine that any scientist or philosopher puts forward is open to question, to question a doctrine is to hold out the possibility that one might conceivably be satisfied, and if one is always going to ask questions, but never wait for an answer or put forward any alternative account, then one's questions degenerate into idle chatter. It is always possible to ask questions, as every small boy discovers around the age of four; but if every question only gives rise to another question, and no answer is ever going to be accepted, then there is not much point in taking the questions seriously—as fathers of small boys around the age of four soon discover.

---

[54] David Hume, *Dialogues on Natural Religion*, Part VIII, *ad fin.* The passage is worth quoting more fully: Each disputant triumphs in his turn, while he carries on an offensive war, and exposes the the absurdities, barbarities, and pernicious tenets of his antagonist. But all of them, on the whole, prepare a complete triumph for the sceptic, who tells them that no system ought ever to be embraced with regard to such subjects: for this plain reason that no absurdity ought ever to be assented to with regard to any subject A total suspense of judgement is here our only reasonable resource. And if every attack, as it is commonly observed, and no defence among theologians is successful, how complete must be his victory who remains always, with all mankind on the offensive, and has himself no fixed station or abiding city which he is ever, on any occasion, obliged to defend?

To the reasonable enquirer, even if not to the incorrigible sceptic, the multiplicity of derivations of the Lorentz transformation
offers a valuable variety of justifications and explanations, capable
of being addressed to different enquirers with different doubts or
difficulties. But besides being dialectically useful, it is philosophically interesting at a different level. It shows the way in which
our concepts are interlocked. Our conceptual scheme is highly connected. The fundamental categories—in this case, time, space and
causality—have many conceptual ties with one another, and are
subject to a variety of conceptual constraints. That time and space
should be causally inefficacious, that time and space should be homogeneous, and that space should be isotropic, are not separate
doctrines which just happen to be true, but ones whose rationale
needs to be teased out carefully. The different derivations of the
Lorentz transformation shows how much more there is at stake,
and how it is that very considerable and precise consequences flow
from fundamental tenets which are not forced on us either by the
law of non-contradiction or by brute empirical fact, but which none
the less rationally commend themselves.

The argument of this section is developed further in §§10.4 and 10.5.

Chapter 6
Four-Vector Transformations

The argument in this chapter follows on that of Chapter 1

§6.1 Spacetime Physics

In Chapter 1 we saw how space and time can be integrated into a single spacetime, in which velocities are a sort of pseudo-angle, and the fundamental magnitude is spacetime separation, $d\sigma$, or equivalently proper time $d\tau$, which is invariant not only under displacement and rotation, but under "pseudo-rotation"—change of velocity—as well. In this chapter we carry on the programme of taking a spacetime view of physics. We need to consider not only spatiotemporal positions and the spatiotemporal separation between them, but other magnitudes such as velocity, momentum, energy, acceleration and force. We want to know what are their relativistic counterparts, and how they change as we transform from one frame of reference to another. There are thus two problems: first we have to define the relativistic counterparts of classical magnitudes; and secondly we have to find out how to transform them, as we go from one inertial frame of reference to another. In resolving the first problem we shall be generalising from three dimensions to four, and seeking "four-vector" equivalents of the quantities of classical physics; in resolving the second, we shall try to make the components of the four-vectors as much like the fundamental spacelike and timelike components of spacetime as possible, so that they also will be covariant under the Lorentz transformation, and have the same "length", that is some scalar quantity that, like spacetime separation, is invariant under the Lorentz transformation.

We have considerable freedom in defining relativistic counterparts of classical magnitudes, the only essential requirement being that they reduce to the corresponding classical ones when the velocity, v, tends to zero, or equivalently when $\beta \to 0$.

Whatever definitions we choose, satisfying this limiting condition, we can work out the corresponding transformation equations and go on to solve whatever problem interests us. If we do this we soon find that in general the transformation equations are complicated, and in physics we always want our equations to be as simple as possible. This not only reduces the chance of errors, but is more likely to display unsuspected fundamental connections. It turns out, as we shall see, that if we form the relativistic magnitudes by simply replacing all derivatives with respect to time by derivatives with respect to the proper time τ, then we can construct four-vectors from the new relativistic magnitudes that transform by the Lorentz transformation. Since the proper time tends to the classical time in the limit of low velocities this method of forming the relativistic counterparts satisfies the limiting condition already mentioned.

In the following sections the proper time is used to define the relativistic velocity and relativistic momentum, and we show that they have the desired properties. We also show that this choice ensures the invariance of mass, and leads to the same transformation equations for acceleration and force. In addition we show that four-vector transformations can also be applied to the components of the electromagnetic field, showing the intimate connection between the Special Theory and Maxwell's equations. Indeed, we recall that it was Einstein's early thoughts about the necessity of ensuring the covariance of Maxwell's equations that led him to the Special Theory. He asked himself what a light wave would look like to an observer moving along with it, and realised that to avoid contradiction it is necessary to use the Lorentz transformation.

§6.2 Relativistic Velocity

The spacetime interval $\mathrm{d}\sigma$ is given by the equation[1]

$$-\mathrm{d}\sigma^2 = c^2\mathrm{d}\tau^2 = c^2dt^2 - dx^2 - dy^2 - dz^2. \qquad (6.2.1)$$

This implies

$$\left(\frac{d\tau}{dt}\right)^2 = 1 - \frac{1}{c^2}\left[\left(\frac{dx}{dt}\right)^2 + \left(\frac{dy}{dt}\right)^2 + \left(\frac{dz}{dt}\right)^2\right] \qquad (6.2.2)$$

and hence

$$\frac{dt}{d\tau} = \left\{1 - \frac{1}{c^2}\left[\left(\frac{dx}{dt}\right)^2 + \left(\frac{dy}{dt}\right)^2 + \left(\frac{dz}{dt}\right)^2\right]\right\}^{-\frac{1}{2}} = \gamma. \qquad (6.2.3)$$

---

[1] See above, §1.5, p.17.

Thus the relativistic velocity in the x-direction is

$$\frac{1}{c}\frac{dx}{d\tau} = \frac{1}{c}\frac{dx}{dt}\frac{dt}{d\tau} = \beta_x\gamma. \tag{6.2.4}$$

To obtain the transformation relations for velocities we start from the Lorentz transformation

$$\begin{pmatrix} \gamma_v & i\beta_v\gamma_v \\ -i\beta_v\gamma_v & \gamma_v \end{pmatrix}\begin{pmatrix} x \\ ict \end{pmatrix} = \begin{pmatrix} x' \\ ict' \end{pmatrix} \tag{6.2.5}$$

where the subscript v refers to the relative velocity of the two frames of reference. Differentiating with respect to τ and dividing by c we obtain

$$\begin{pmatrix} \gamma_v & i\beta_v\gamma_v \\ -i\beta_v\gamma_v & \gamma_v \end{pmatrix}\begin{pmatrix} \gamma\beta_x \\ i\gamma \end{pmatrix} = \begin{pmatrix} \gamma'\beta'_x \\ i\gamma' \end{pmatrix} \tag{6.2.6}$$

which is the transformation equation for velocities along the x-axis. Restoring the y and z components we obtain the *velocity four-vector* and its corresponding transformation equation

$$\begin{pmatrix} \gamma_v & 0 & 0 & i\beta_v\gamma_v \\ 0 & 1 & 0 & 0 \\ 0 & 0 & 1 & 0 \\ -i\beta_v\gamma_v & 0 & 0 & \gamma_v \end{pmatrix}\begin{pmatrix} \gamma\beta_x \\ \gamma\beta_y \\ \gamma\beta_z \\ i\gamma \end{pmatrix} = \begin{pmatrix} \gamma'\beta'_x \\ \gamma'\beta'_y \\ \gamma'\beta'_z \\ i\gamma' \end{pmatrix} \tag{6.2.7}$$

We note that the "length" of the vector is constant. The relation (6.2.6) implies that

$$\gamma_v\gamma\beta_x - \beta_v\gamma_v\gamma = \gamma'\beta'_x \quad \text{and} \quad -\beta_v\gamma_v\gamma\beta_x + \gamma_v\gamma = \gamma'. \tag{6.2.8}$$

Eliminating γ' gives

$$\beta'_x = \frac{\beta_x - \beta_v}{1 - \beta_x\beta_v} \tag{6.2.9}$$

which is the velocity addition law already obtained in Chapter 1.[2] The corresponding relations for the y and z components are

$$\beta'_y = \frac{\beta_y}{\gamma_v(1 - \beta_x\beta_v)} \quad \text{and} \quad \gamma_v\beta'_z = \frac{\beta_z}{\gamma_v(1 - \beta_x\beta_v)} \tag{6.2.10}$$

---

[2] §1.2

(These are not symmetric in x, y and z, because the motion is along the x-axis.)

Thus by defining relativistic velocity as a derivative with respect to the proper time we obtain a four-vector velocity that transforms according to the Lorentz transformation. If we had kept to the definition of velocity as a derivative with respect to the Newtonian time we would still have the velocity addition law but the components would no longer transform by the Lorentz transformation.

§6.3 Relativistic Momentum

In Newtonian mechanics the momentum is defined as the product of the mass and the velocity. In relativistic mechanics we retain this definition, but instead of the Newtonian velocity dx/dt we use the relativistic velocity $dx/d\tau$. Thus the transformation equation for momentum is simply (6.2.6) multiplied by m,

$$\begin{pmatrix} \gamma_v & i\beta_v\gamma_v \\ -i\beta_v\gamma_v & \gamma_v \end{pmatrix} \begin{pmatrix} m\beta_x\gamma \\ im\gamma \end{pmatrix} = \begin{pmatrix} m\beta_x'\gamma' \\ im\gamma' \end{pmatrix} \tag{6.3.1}$$

What is the physical meaning of the component $m\gamma$? Let us express γ in terms of β and expand it:

$$m\gamma = m(1 - \beta^2)^{-\frac{1}{2}}$$

and, as v tends to 0,

$$m(1 - \beta^2)^{-\frac{1}{2}} \text{ tends to } m + \frac{1}{2}m\beta^2 \tag{6.3.2}$$

Thus in the limit of low velocities $m\gamma$ is the sum of the mass m and the kinetic energy $\frac{1}{2}m\beta^2$, in units of c^2. If now we now regard the mass as having an energy associated with it, the so-called rest energy, we can define the total (relativistic) energy

$$E = m\gamma \tag{6.3.3}$$

Putting $p = m\beta\gamma$ for the relativistic momentum and $E = m\gamma$ for the relativistic energy gives the transformation equation for energy and momentum

$$\begin{pmatrix} \gamma_v & i\beta_v\gamma_v \\ -i\beta_v\gamma_v & \gamma_v \end{pmatrix} \begin{pmatrix} p_x \\ iE \end{pmatrix} = \begin{pmatrix} p_x' \\ iE' \end{pmatrix} \tag{6.3.4}$$

Restoring the y and z components as before gives the *energy-momentum four-vector*, which transforms according to the Lorentz transformation

$$
\begin{pmatrix}
\gamma_v & 0 & 0 & i\beta_v\gamma_v \\
0 & 1 & 0 & 0 \\
0 & 0 & 1 & 0 \\
-i\beta_v\gamma_v & 0 & 0 & \gamma_v
\end{pmatrix}
\begin{pmatrix}
p_x \\ p_y \\ p_z \\ iE
\end{pmatrix}
=
\begin{pmatrix}
p'_x \\ p'_y \\ p'_z \\ iE'
\end{pmatrix}
\tag{6.3.5}
$$

Once again, the length of the four-vector is a constant. Thus

$$
E^2 = p^2 + m^2 \tag{6.3.6}
$$

as indeed follows from the definitions of E, p and γ. The same expression for the relativistic energy was obtained in a different way in §1.5.

§6.4 Mass-Energy

The association of mass and energy mentioned in the last section has profound consequences. Expressed in more familiar units, the total energy in the limit of low velocities is $mc^2 + \frac{1}{2}mv^2$, the rest energy mc^2 plus the kinetic energy $\frac{1}{2}mv^2$. Thus if the total energy is conserved, then the theory of relativity tells us that mass is equivalent to energy and gives us the equation $E_0 = mc^2$ from which we can calculate the energy E_0 equivalent to a mass m.

In the above discussion the mass has been treated as a constant. If however we had kept the Newtonian definition of velocity, and also insisted that the velocity transforms by the Lorentz transformation, then we would be forced to allow the mass to vary with the velocity according to the relation $m = m_0\gamma$, where now m_0 is an invariant quantity.

This can be expressed in another way by noting that the relativistic momentum is $m\beta\gamma$. We then have to choose between defining the mass as m and the velocity as $\beta\gamma$, or the mass as $m\gamma$ and the velocity as β. It is repugnant to have the mass depending on velocity, whereas we know that velocities behave in a non-Newtonian way, so it is very natural to make the former choice, and this has the additional advantage that the velocities so defined transform by the Lorentz transformation. Thus we see that by defining the velocity as a derivative with respect to the proper time allows the mass to remain an invariant quantity.

It is very often said in textbooks that the theory of relativity has shown that mass increases with velocity according to the relation $m = m_0\gamma$, where m_0 is the rest mass. Expressed in this way it appears as a new and profound property of matter, whereas it is really a result of a particular definition of relativistic velocity. If we insist on retaining Newtonian dynamics, and the Newtonian definitions of velocity and acceleration, then we can still obtain relativistically correct results if we pay the price of allowing the mass to depend on the velocity. If however we adopt Einsteinian dynamics, the mass remains invariant.

This device of including some dynamical aspect of a problem by allowing the mass to depend on another variable is often used in other branches of physics. Thus an electron in a crystal is subject to the fields of the nearby atoms, and this affects its trajectory. If we know these fields we can calculate its motion. Another and completely equivalent way of describing the motion is to take account of the effects of the atomic fields by allowing the mass of the electron to vary with its position in the lattice.

A second example is provided by the motion of a particle in a momentum-dependent potential. The Hamiltonian is then

$$H = T + V = \frac{p^2}{2m} + V_0 + V_1 p^2 \qquad (6.4.1)$$

which may be written

$$H = \frac{p^2}{2m^*} + V_0 \qquad (6.4.2)$$

where the effective mass m^* is given by

$$\frac{1}{2m^*} = \frac{1}{2m} + V_1 \qquad (6.4.3)$$

Thus the motion of a particle of mass m in a momentum-dependent potential is exactly the same as that of a particle of mass m^* in a momentum-independent potential. This example shows clearly that the concept of effective mass is simply a convenient mathematical device, devoid of profound philosophical import.

But if allowing the mass to depend on velocity is just a mathematical device, does the same not apply to the concept of relativistic energy and hence to the equivalence of mass and energy? This is a question that can be answered by experiment. Einstein showed that if a body emits or absorbs radiation then its mass changes at the same time according to $E = mc^2$. A particular example of the effect of changes of mass on the dynamics of collisions is described in §6.5. These changes in mass can be measured, and Einstein's relation for the equivalence of mass and energy has been confirmed to high precision by measurements of the energy released in nuclear reactions. A more spectacular demonstration of the equivalence of mass and energy was provided at Alamogordo.

§6.5   Conservation of Energy and Momentum

The laws of conservation of energy and of momentum are of great importance in Newtonian mechanics, and we want to ensure that our relativistic definitions preserve them. We will now show by a simple example that this is indeed the case, providing we substitute the conservation of total energy (which for small velocities becomes the rest energy plus the kinetic energy or mass energy) for the Newtonian conservation of kinetic energy.

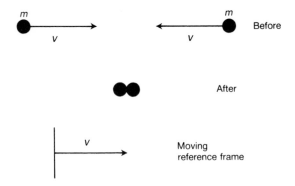

Figure 6.5.1   Two particles of equal mass m collide with each other, and stick together.

Consider the collision of two particles of equal mass m moving towards each other along the same straight line with equal velocities v. After the collision they stick together and are of course at rest. Now consider the collision from the reference frame of the left-hand particle. Before the collision this particle is at rest and the other particle is moving towards it at a velocity $v'=c\beta'$, where from (6.2.9)

$$\beta' = \frac{2\beta}{1 + \beta^2} \qquad (6.5.1)$$

After the collision the composite system of mass $2m$ is moving relative to the moving frame with a velocity β. Thus in this frame the total momentum before the collision is $m\beta'\gamma'$, and that after the collision would seem to be $2m\beta\gamma$.

Now

$$m\beta'\gamma' = \frac{2m\beta}{1 + \beta^2} \cdot \frac{1}{\left(1 - \left(\frac{2\beta}{1+\beta^2}\right)^2\right)^{\frac{1}{2}}} = \frac{2m\beta}{1 + \beta^2} \cdot \frac{1 + \beta^2}{1 - \beta^2} = 2m\beta\gamma^2$$

$$(6.5.2)$$

which is not the same as $2m\beta\gamma$. It thus seems that momentum is conserved in the rest frame but not in the moving frame. What has gone wrong?

What we have forgotten is that the kinetic energy of the two particles before the collision has been converted into extra mass that must be added to that of the two particles at rest. This extra mass is that equivalent to the total energy less the rest energy $(m\gamma - m)$ for each particle. Thus instead of $2m$ we have $2m\gamma$ in the expression for the momentum after the collision, it thus becomes $2m\beta\gamma^2$, in agreement with the momentum before the collision. Thus we obtain a consistent account of the collision, satisfying the laws of conservation of momentum and mass-energy, and valid in all uniformly-moving reference frames. Note that it would not be correct just to add the mass equivalent of the classical kinetic energy, namely $2 \times \frac{1}{2}mv^2 = m\beta^2$ to the rest mass $2m$. This would be equivalent to assuming that (6.3.2) is exactly true, whereas it is only correct in the limit of small velocities. The total energy is $m\gamma$, not $m + \frac{1}{2}m\beta^2$.

This example shows how the relativistic generalisations of momentum and energy ensure that the conservation laws retain their validity in a Lorentz-invariant way. One example is not enough to show that this is always the case but a general proof can be constructed and may be found in standard textbooks.

A further differentiation of equation (6.2.5) with respect to the proper time and multiplication by the invariant mass gives the transformation equation for force

$$
\begin{pmatrix}
\gamma_v & 0 & 0 & i\beta_v\gamma_v \\
0 & 1 & 0 & 0 \\
0 & 0 & 1 & 0 \\
-i\beta_v\gamma_v & 0 & 0 & \gamma_v
\end{pmatrix}
\begin{pmatrix}
F_1 \\ F_2 \\ F_3 \\ iF_4
\end{pmatrix}
=
\begin{pmatrix}
F_1' \\ F_2' \\ F_3' \\ iF_4'
\end{pmatrix}
\tag{6.5.3}
$$

where (F_1, F_2, F_3, iF_4) is the four-vector force. Division by m gives the transformation equation for acceleration.

§6.6   The Integration of Electricity and Magnetism

The spacetime approach can be applied to electromagnetism as well as to Newtonian mechanics, and provides profound insights into the nature of magnetism. As before, we try to generalise from three-vectors to four-vectors in such a way that they will be Lorentz-covariant. Often this means modifying previously established laws, but this is nothing new. We are continually recasting and altering the formulation of natural laws so as to put them in a more satisfactory form from a theoretical point of view. Coulomb's law of electrostatic repulsion—that the repulsion (or attraction) between two similarly (or oppositely) charged bodies is proportional to the product of the charges and inversely proportional to the square of the distance between them—is entirely satisfactory so far as experimental evidence goes, but involves action at a distance.[3] It is better to posit a spatiotemporal field, and express Coulomb's law in terms of the field instead of an inexplicable force. In the same spirit we generalise from points to volumes. Instead of just the field generated by a point source, we consider some volume enclosed by a surface, and say that the flux of the field through that surface is proportional to the electric charge within it. That is, instead of a force law that the repulsive force between two charges, e_1 and e_2 separated by a distance r is

$$
k\frac{e_1 \times e_2}{r^2}.
\tag{6.6.1}
$$

---

[3] See above, ch.1, §1.1, and below §10.5.

we talk in terms of the divergence of a field, and say

$$div \ \mathbf{E} \equiv \nabla.\mathbf{E} = \rho. \tag{6.6.2}$$

In this case we have altered the way we express a natural law. Sometimes we modify it by adding an extra term as we seek greater generality; thus Ampère's equation, that the current is equal to the curl of the magnetic field,

$$curl \ \mathbf{B} \equiv \nabla \times \mathbf{B} = j, \tag{6.6.3}$$

holds only for steady current: so we add a term $\frac{\partial \mathbf{E}}{\partial t}$ to take account of changes of the electric field $\mathbf{E}$, and have, instead of (6.6.3),

$$\nabla \times \mathbf{B} = \mathbf{j} + \frac{\partial \mathbf{E}}{\partial t}. \tag{6.6.4}$$

This is typical. Ampère discovers a law. It is confirmed experimentally. We think we are on to something. But it is the wrong "shape" to hold under some conditions. So we add on a bit, to make it the right shape to hold under these conditions too. We think there must be some law, of which we have found only a special case, and so reconstruct the special case to be of a general form that can hold in all circumstances. We are impelled not simply by experimental evidence but also by considerations of symmetry and elegance. In the same way, only much more explicitly, we impose the condition of being covariant under the Lorentz transformation as one that any properly formulated law of the Special Theory must satisfy.

To express electric and magnetic fields in four-vector form we begin by making a crucial assumption: that the electric charge is invariant under the Lorentz transformation, and is conserved over time. If this assumption be granted, then for any arbitrary volume V bounded by a surface S,

$$\int_V \frac{\partial \rho}{\partial t} dV + \int_S \mathbf{j}.\mathbf{dS} = 0, \tag{6.6.5}$$

where $\mathbf{j}$ is the electric current.

Gauss' theorem tells us that the second term is equal to the change of electric charge throughout the whole volume, that is

$$\int_S \mathbf{j}.\mathbf{dS} = \int_V \nabla.\mathbf{j} dV, \tag{6.6.6}$$

so that

$$\int_V \frac{\partial \rho}{\partial t} dV + \int_V \nabla . \mathbf{j} \, dV = 0. \qquad (6.6.7)$$

Thus for an arbitrarily small volume we obtain the continuity or charge conservation condition

$$\frac{\partial \rho}{\partial t} + \nabla . \mathbf{j} = 0. \qquad (6.6.8)$$

The form is suggestive: $\mathbf{j}$ is a three-vector, (j_x, j_y, j_z), and $\nabla . \mathbf{j}$ stands for $\frac{\partial j_x}{\partial x} + \frac{\partial j_y}{\partial y} + \frac{\partial j_z}{\partial z}$, and is therefore invariant under displacement and rotation in three dimensions. Can we find a corresponding four-vector in terms of (x, y, z, ict) in line with our general programme of §6.1? The corresponding four-vector operator will be

$$\left(\frac{\partial j_x}{\partial x}, \frac{\partial j_y}{\partial y}, \frac{\partial j_z}{\partial z}, \frac{\partial ic\rho}{\partial ict} \right)$$

so we can rewrite 6.6.8) as

$$\frac{\partial j_x}{\partial x} + \frac{\partial j_y}{\partial y} + \frac{\partial j_z}{\partial z} + \frac{\partial ic\rho}{\partial ict} = 0, \qquad (6.6.9)$$

which we could express as the four-vector divergence of $(j_x, j_y, j_z, ic\rho)$ being equated to zero. (Often ρ is chosen so that $c = 1$; we can then write this four-vector as $(\mathbf{j}, i\rho)$. We have thus expressed the conservation of charge in four-vector form, making use of the field version of Coulomb's law.

In our search for simplicity we have constructed a new four-vector $(j_x, j_y, j_z, ic\rho)$, which we may write as $\mathbf{J}$ with components (now putting the time-like one first) J_1, J_2, J_3, J_4 which we write with a Greek subscript, μ, as J_μ. What does $\mathbf{J}$ represent? It represents a combination of the electrodynamic current density in each direction and the electrostatic charge density. It has a good chance of representing something real. If it does, it is something that is changing over time and over space, and we may wonder how such a change may be propagated. The most explicable, though not the only conceivable, way of propagating causal influence is by means of waves. It is natural to consider whether the $\mathbf{J}$ might not itself be the waves of some more *recherché* field, $\mathbf{A}$. If that were so, it would be necessary for each $\mu = 1, 2, 3, 4$, that

$$\frac{\partial^2 A_\mu}{c^2 \partial t^2} - \frac{\partial^2 A_\mu}{\partial x^2} - \frac{\partial^2 A_\mu}{\partial y^2} - \frac{\partial^2 A_\mu}{\partial z^2} = J_\mu$$

or, using the D'Alembertian operator, $\Box^2 \equiv \frac{\partial^2}{c^2 \partial t^2} - \nabla^2$

$$\Box^2 \mathbf{A}_\mu = J_\mu. \tag{6.6.10}$$

The four-vector $\mathbf{A}$ will, like $\mathbf{J}$, consist of three spacelike and one timelike components, A_x, A_y, A_z, icV.[4] It is, however, much more rarefied than $\mathbf{J}$. Indeed, it is not completely specified by the condition given above

$$\Box^2 A_\mu = J_\mu$$

and we can alter the timelike component icV, provided we make suitable further alterations to the spacelike components A_x, A_y, A_z, and *vice versa*. In general, we can add to $\mathbf{A}$ the four-vector gradient of any arbitrary potential field χ and leave (6.6.10) unchanged. That is, if we make the transformation $\mathbf{A}' = \mathbf{A} + \nabla\chi$, $V' = V - \frac{\partial \chi}{\partial t}$, then

$$\Box^2 A'_\mu = J_\mu. \tag{6.6.11}$$

We therefore add the further "gauge condition"

$$\frac{\partial A_x}{\partial x} + \frac{\partial A_y}{\partial y} + \frac{\partial A_z}{\partial z} + \frac{\partial icV}{\partial ict} = 0. \tag{6.6.12}$$

which is analogous to the four-vector divergence expression of Coulomb's law for $\mathbf{J}$. The gauge condition can be expressed more briefly as

$$\Box \mathbf{A} = 0 \quad \text{or} \quad \nabla \cdot \mathbf{A} + \frac{\partial V}{\partial t} = 0. \tag{6.6.13}$$

If $\mathbf{A}$ represents a real field, its vector derivatives must be of importance too. But they are not as simple as with three-dimensional fields. In three dimensions the curl of a field

$$\nabla \times \mathbf{A} = \left(\frac{\partial A_z}{\partial y} - \frac{\partial A_y}{\partial z}, \frac{\partial A_x}{\partial z} - \frac{\partial A_z}{\partial x}, \frac{\partial A_y}{\partial x} - \frac{\partial A_x}{\partial y} \right) \tag{6.6.14}$$

---

[4] It is easy to be confused between three-vectors and four-vectors. In this passage the four-vector $\mathbf{A}$, and its four-vector components A_μ are printed in bigger type, while the three spacelike components A_x, A_y, A_z, which together form a three-vector, A, are printed normally, as also the timelike component V. Thus $A = A_\mu = (A_0, A_1, A_2, A_3) = (icV, A_x, A_y, A_z) = (icV, \mathbf{A})$ also written $(\mathbf{A}, icV)$

is a simple vector; but when we generalise to more dimensions, it ceases to be a simple vector, and becomes a second-order tensor instead. With four dimensions, the tensor is a two-dimensional array, $\mathbf{\Phi}$, where

$$\Phi_{\mu\nu} = \frac{\partial A_\nu}{\partial x_\mu} - \frac{\partial A_\mu}{\partial x_\nu}. \tag{6.6.15}$$

This can be expressed in matrix form as a 4×4 antisymmetric matrix:

$$\begin{pmatrix} 0 & \Phi_{12} & \Phi_{13} & \Phi_{14} \\ -\Phi_{12} & 0 & \Phi_{23} & \Phi_{24} \\ -\Phi_{13} & -\Phi_{23} & 0 & \Phi_{34} \\ -\Phi_{14} & -\Phi_{24} & -\Phi_{34} & 0 \end{pmatrix}. \tag{6.6.16}$$

We want this tensor to represent the electromagnetic field, so its elements must be functions of the components of the magnetic and electric fields. We now show that if we identify the elements of the tensor with the components of the fields in a particular way, then Maxwell's equations follow as a logical consequence.

First we identify

$$B_x = \frac{\partial A_z}{\partial y} - \frac{\partial A_y}{\partial z} \equiv \Phi_{23}$$

$$B_y = \frac{\partial A_x}{\partial z} - \frac{\partial A_z}{\partial x} \equiv -\Phi_{13} \tag{6.6.17}$$

$$B_z = \frac{\partial A_y}{\partial x} - \frac{\partial A_x}{\partial y} \equiv \Phi_{12}$$

so that

$$\mathbf{B} = \nabla \times \mathbf{A}, \tag{6.6.18}$$

and therefore

$$\nabla \cdot \mathbf{B} = 0. \tag{6.6.19}$$

Secondly we identify

$$-\frac{iE_x}{c} = \frac{\partial A_4}{\partial x} - \frac{\partial A_x}{i\partial t} \equiv \Phi_{14}$$

$$-\frac{iE_y}{c} = \frac{\partial A_4}{\partial y} - \frac{\partial A_y}{i\partial t} \equiv \Phi_{24}. \tag{6.6.20}$$

$$-\frac{iE_z}{c} = \frac{\partial A_4}{\partial z} - \frac{\partial A_z}{i\partial t} \equiv \Phi_{34}$$

The fourth component of **A**

$$A_4 \equiv \frac{iV}{c} \tag{6.6.21}$$

so that

$$
\begin{aligned}
E_x &= -\frac{\partial V}{\partial x} - \frac{\partial A_x}{\partial t} \\
E_y &= -\frac{\partial V}{\partial x} - \frac{\partial A_y}{\partial t} \\
E_z &= -\frac{\partial V}{\partial x} - \frac{\partial A_z}{\partial t}
\end{aligned} \tag{6.6.22}
$$

Thus

$$\mathbf{E} = -\frac{\partial \mathbf{A}}{\partial t} - \nabla V \tag{6.6.23}$$

Therefore

$$\nabla \times \mathbf{E} = -\frac{\partial}{\partial t} \nabla \times \mathbf{A} = -\frac{\partial \mathbf{B}}{\partial t} \tag{6.6.24}$$

which is the Faraday-Lenz law.

Furthermore

$$
\begin{aligned}
\nabla \times \mathbf{B} &= \nabla \times \nabla \times \mathbf{A} \\
&= \nabla \nabla \cdot \mathbf{A} - \nabla^2 \mathbf{A} \\
&= \nabla(-\frac{\partial V}{\partial t}) + \mathbf{j} - \frac{\partial^2 \mathbf{A}}{\partial t^2} \quad \text{by (6.6.13) and (6.6.10)} \\
&= \nabla(-\frac{\partial V}{\partial t}) + \mathbf{j} + \frac{\partial \mathbf{E}}{\partial t} + \frac{\partial}{\partial t}(\nabla V) \quad \text{by (6.6.23)}
\end{aligned}
$$

Thus

$$\nabla \times \mathbf{B} = \mathbf{j} + \frac{\partial \mathbf{E}}{\partial t} \tag{6.6.4}$$

which is Ampère's law, in its modified form.

The equations (6.6.2), (6.6.24), (6.6.19) and (6.6.4) together are Maxwell's equations; for ease of convenience we collect them together in a box.

$$
\boxed{
\begin{aligned}
\nabla \cdot \mathbf{E} &= \rho & \text{(Gauss)} \\
\nabla \times \mathbf{E} &= -\frac{\partial \mathbf{B}}{\partial t} & \text{(Faraday-Lenz)} \\
\nabla \cdot \mathbf{B} &= 0 & \text{(No magnetic monopole)} \\
\nabla \times \mathbf{B} &= \mathbf{j} + \frac{\partial \mathbf{E}}{\partial t} & \text{(Modified Ampère)}
\end{aligned}
}
$$

The electromagnetic field tensor has the form

$$\Phi = \begin{pmatrix} 0 & B_z & -B_y & -iE_x/c \\ -B_z & 0 & B_x & -iE_y/c \\ B_y & -B_x & 0 & -iE_z/c \\ iE_x/c & iE_y/c & iE_z/c & 0 \end{pmatrix} \qquad (6.6.25)$$

Since it is covariant, it may be transformed from one frame of reference to another by

$$f'_{\mu\nu} = \mathcal{L}\Phi_{\mu\nu}\mathcal{L}', \qquad (6.6.26)$$

where $\mathcal{L}$ is the Lorentz transformation.

The intimate relation between electricity and magnetism is shown by the transformation properties of the electromagnetic field tensor. Consider a field $\mathbf{E}$ due to a static charge distribution, with no magnetic field. The electromagnetic field tensor is

$$f = \begin{pmatrix} 0 & 0 & 0 & -iE_x/c \\ 0 & 0 & 0 & -iE_y/c \\ 0 & 0 & 0 & -iE_z/c \\ iE_x/c & iE_y/c & iE_z/c & 0 \end{pmatrix} \qquad (6.6.27)$$

Now transform to another frame of reference moving with uniform velocity, say in the Z direction, with respect to the first frame of reference. The Lorentz transformation yields

$$f' = \begin{pmatrix} 0 & 0 & \beta\gamma E_x/c & -i\gamma E_x/c \\ 0 & 0 & \beta\gamma E_y/c & -i\gamma E_y/c \\ -\beta\gamma E_x/c & -\beta\gamma E_y/c & 0 & -iE_z/c \\ i\gamma E_x/c & i\gamma E_y/c & iE_z/c & 0 \end{pmatrix} \qquad (6.6.28)$$

Thus

$$\begin{aligned} cB'_x &= \beta\gamma E_y & E'_x &= \gamma E_x \\ cB'_y &= -\beta\gamma E_x & E'_y &= \gamma E_y \\ cB'_z &= 0 & E'_z &= E_z \end{aligned} \qquad (6.6.29)$$

The transformation shows that there is a magnetic field in the moving frame of reference, although there was none in the first frame of reference. Thus the magnetic field appears as an effect of the transformation from the one frame of reference to the other.

The electromagnetic field tensor provides a powerful way of calculating many electromagnetic phenomena. For example, if we wish to calculate the electromagnetic field of a moving electron, we first write down the electromagnetic field tensor in the frame of reference of the electron itself; this has $\mathbf{B} = 0$ and the electric field given by Coulomb's law, $\mathbf{E} = (e/r^2)$ ř (where ř is a unit vector in the direction of $\mathbf{r}$). Applying the Lorentz transformation yields the electromagnetic field in the observer's frame of reference. Thus from Coulomb's law and the Lorentz transformation we can deduce the fields of a moving charge, and this is equivalent to a complete specification of electrodynamics, encapsulated in Maxwell's equations. But we can still ask if these are really all the assumptions we need, or whether there are other hidden assumptions which we should make explicit and consider in more detail. This will be done in the next section.

§6.7 Derivation of Maxwell's Equations

The reasoning of the previous section is loose. It is an exercise in the optative mood: wouldn't it be convenient, we say, if electric and magnetic phenomena could all be described in terms of the differential geometry of fields, if they were Lorentz invariant, if they can be expressed as the D'Alembertian of other deeper fields, the curl of which was a 4 × 4 tensor, with components we could identify with those turning up in Maxwell's equations. It is almost too good to be true that this should all be the case, and we are rightly suspicious of any argument that it must be so.

The reasoning can be made more rigorous. Purcell gives gives a closely argued derivation in full. He claims:

> From our present vantage point, the magnetic interaction of electric currents can be seen as the inevitable corollary of Coulomb's law. If the postulates of relativity are valid, if electric charge is invariant, and if Coulomb's law holds, then the effects we commonly call magnetic are bound to occur. They will emerge as soon as we examine the electric interaction between a moving charge and other moving charges.[5]

---

[5] Edgar M. Purcell, *Electricity and Magnetism*, New York, 1963, p.173. For a brief discussion, see Robert Resnik, *Introduction to Special Relativity*, New York, 1968, §4.6, pp.176-177.

Besides the premiss that the laws of physics are Lorentz-covariant, we have to assume three "electrostatic" premisses, that electrostatic attraction is governed by an inverse square law, and that electric charge is conserved and that it is Lorentz-invariant. Inverse square laws are very suitable in a three-dimensional space, and have other merits that make them rationally attractive.[6] The conservation and Lorentz-invariance of electric charge is less self-evident. It needs, in any case, to be distinguished from the *conservation* of charge, which is, like the conservation of energy and of mass in Newtonian mechanics, a sameness over time. Lorentz-invariance is not a sameness over time but over any Lorentz transformation from one inertial frame to some other one. Where we have some transformation under which some form is covariant, it is natural to expect that there should be also some feature that is invariant. In quantum mechanics we often find first a group of transformations and seek some group-invariant, and in the eighteenth century Kant argued for the conservation of mass on *a priori* grounds.[7] His arguments as propounded establish less than he claims, but he seems to be feeling towards some principle of persistence and continuing identity, and so of some things having to remain the same whenever there are differences.[8]

How far can we appropriate such arguments for conservation and the Lorentz-invariance of electric charge? The answer, as often, is that we can do so to some extent, but cannot expect them to take us all the way. Once we have recognised the fundamental importance of the Lorentz transformations, we should expect there to be some Lorentz-invariance. But we have no *a priori* grounds for identifying it with the electrostatic phenomena of ordinary experience—sparks when we take a nylon shirt off in dry weather, or the hair on one's hands and arms being made to stand on end by rubbed amber. At the most generous, we might concede that there ought to be some Lorentz-invariant quantity, and that *if* there is any force between two such quantities, it should obey the inverse square law. But

---

[6] See further below, §10.5.

[7] I.Kant, *Critique of Pure Reason*, First Analogy, A182/B224 ff.

[8] P.F. Strawson, *The Bounds of Sense*, London, 1966, ch.3, §3, esp. pp.125-132; W.H.Walsh, *Kant's Criticism of Metaphysics*, Edinburgh, 1975, §24, pp.129-135.

we can offer, at least so far, no reason why there should be such a force, nor why it should be repulsive rather than, as with gravitation, attractive. These are still contingent conditions, vindicated by empirical observation but not, as yet, explicable by reason. To that extent the derivation of Maxwell's equations from the Lorentz transformation is still incomplete.[9]

There are still other assumptions which a critic could call in question. As Feynman says

It is sometimes said, by people who are careless, that all of electrodynamics can be deduced solely from the Lorentz transformation and Coulomb's law. Of course, that is completely false. First, we have to suppose that there is a scalar potential and a vector potential that together make a four-vector. That tells us how the potentials transform. Then why is it that the effects at the retarded time are the only things that count? Better yet, why is it that the potentials depend only on the position and the velocity and not, for instance, on the acceleration. The *fields* **E** and **B** *do* depend on the acceleration. If you try to make the same kind of an argument with respect to them, you would say that they depend upon the position and velocity at the retarded time. But then the fields from accelerating charge would be the same as the fields from a charge at the projected position—which is false. The *fields* depend not only on the position and the velocity along the path but also on the acceleration. So there are several additional tacit assumptions in this great statement that everything can be deduced from Lorentz transformation. (Whenever you see a sweeping statement that a tremendous amount can come from a very small number of assumptions, you always find that it is false. There are usually a large number of implied assumptions that are far from obvious if you think about them sufficiently carefully.)[10]

---

[9] W.Rindler, *Essential Relativity*, New York, 2nd ed. 1977,, Appendix II, pp.247-51, 'How to "Invent" Maxwell's Theory': Rindler assumes four-potential, and also that (i) Field of force is *linear* function of velocity of particle, (ii) it leaves rest mass unaltered, (iii) charge is independent of velocity, (iv) charge is conserved.

[10] *The Feynman Lectures*, vol 2. §26-1.

That these, and many other, assumptions are made is not in question. What is to be questioned is whether exception should be taken to making some such assumptions as these. Physicists are in the habit of assuming a lot which is, strictly, not fully established. They assume that space is isotropic and homogeneous, that the parallelogram of forces is correct, that two velocities in the same direction when compounded will be in the same direction still. If they did not make some such assumptions, physics would be impossible. We must make strong assumptions of irrelevance in order to generalise from particular observations, discounting many features in order to concentrate on some. We may occasionally be wrong, as Oersted was.[11] If we are, we have to think again, maybe finding some mistake in our procedure, more likely finding some further factor whose irrelevance we had overlooked, but only very occasionally altogether rejecting the assumption we had made.

Purcell offers a very interesting programme. Historically the Lorentz transformation was derived as that under which Maxwell's equations were covariant. But the Lorentz transformation, we have seen, can be derived in many other ways, and therefore it is natural to ask whether the historical derivation can be made to go the other way, and yield Maxwell's equations from the Lorentz transformation. The answer is a qualified Yes. Other assumptions are required too, just as they are for the different derivations of the Lorentz transformation. They may be contested, and it is right to question them. But if they are granted, we have not only a new direction of approach to Maxwell's equations, but a radically different view of physics.

The argument of this section is continued in §§10.2, 10.3, 10.5 and 10.7.

---

[11] See below, §10.7.

Chapter 7
Counterparts, Parity,
Conservation and Symmetry

The argument in this chapter
follows on from that of Chapter 2

§7.1 Counterparts

The controversy between Clarke and Leibniz ended in a draw. Leibniz had maintained that space and time were not real things, but only ideal things, indeed only a system of relations. He was able to argue effectively that since according to the canons of scientific thought it makes no difference to the laws of nature where or when an experiment is carried out, we cannot determine absolutely where or when anything is, but only in relation to some other thing or event. Clarke detected some internal incoherence in Leibniz's position, but was unable to convict him of actual inconsistency before Leibniz's death terminated the correspondence. He was able to make the point that Newtonian mechanics distinguishes absolute accelerations from uniform motions,[1] but was not able to gainsay Leibniz's claim that everything we can say about the spatial or temporal properties of a thing can be naturally expressed in relational terms.

Kant hit upon a new argument. He pointed out that a relational description of a right hand and a left hand are the same, whereas in fact they are evidently different. There is therefore something—handedness we might call it—which is a spatial property but which is not expressible in relational terms. He discussed the problem of "Incongruous Counterparts", as he called it, in a paper he published in 1768 under the title "Concerning the Ultimate Foundation of the Differentiation of Regions in Space".[2] He regarded it then as constituting a decisive refutation of Leibniz, and proving the reality of space; later he changed his mind and

---

[1] See §1.1 p.2, n.2, and §8.4, n.10.

[2] Translated in G.B.Kerford and D.R.Walford, *Kant: Selected Precritical Writings*, Manchester, 1968, pp.36-43.

in his *Prolegomena*[3] argued that the difference could not reside in
the things themselves, but must be accounted for by their relation
to our sensibility. His second thoughts were less happy than his
first, and subsequent discussion has been clouded by a confusion
of the two questions (1) whether there is any difference between a
right hand and a left hand, and (2) how we could tell which was
which. It is reasonable to maintain that the answer to the second
question must turn on some relation to us: after all it is we who
are being asked to tell which is which. If there were nothing else
in the universe except a left hand and a right hand, we might well
be unable to say that one was the left hand; only if we were there
with a body, in which the heart was on the left hand side or with a
natural preference for using the right hand, could we identify either
hand as being left. But it does not at all follow that, in the absence
of some corporeal being, we could not tell that there was a difference. One can imagine oneself, divorced from one's body, located
in an otherwise empty universe observing two hands or gloves or
helixes, and seeing that they could not be brought by any continuous motion to be superimposed one on the other, and so that
they were different even though the internal relations of the parts
of each was the same as that of the parts of the other. Indeed, it is
not stretching the imagination too much to think of oneself as possessing some powers of psychokinesis, and being able to will that
either object moves as desired, and then discovering that there was
no way of bringing them to coincide. There is a clear difference
between a characterization in terms of distances and angles, and a
characterization in terms of continuous motions. The former are
invariant under the whole Euclidean group, the latter under only
the *proper* Euclidean group. Once that point is taken the verificationist argument for there being no real difference loses its charm.
We see that there is a conceptual difference between a left-handed
configuration and a right-handed configuration, in that neither can
be transformed into the other by a transformation of the proper
Euclidean group even though we cannot say which is which. If
the only things in the universe were two left hands, we might well
be unable to identify them as left hands rather than right hands,
but we could see that they were the same as regards handedness,
whereas if there were one left and one right hand, then we could

---

[3] Translated by P.G.Lucas, Manchester, 1953, §13, pp.41-43.

see that they wére different. So there is something more to space than the relations between objects, and a purely relational account of space, apart from any other defects, is clearly defective on this score.

Nerlich identifies what has been left out as a topological feature, the "orientability" of space.[4] Most spaces are orientable, and do not admit of any continuous transformation of a left-handed figure into a right-handed one: a conspicuous exception being a Möbius strip, made by twisting a length of tape by half a turn and gumming the opposite ends together, on which a two-dimensional figure can be slid round until it coincides with its mirror image. Our space, we know, is locally orientable—no continuous motion in our locality can—and, so far as we know, it is globally orientable too. In order to characterize our space adequately, we need to specify not only the spatial relations of things in space, but the topological proper-ties of space as a whole: we need to say that it is orientable, and also that it is continuous, that it is three-dimensional, that it is unbounded, and that it is simply connected. These are properties not of this or that thing, but of space as a whole. It is logically correct to express them by predicating of 'Space', a substantive, perhaps dignified with a capital initial letter, the various topologi-cal properties that it has: which is tantamount to saying that Space is a substance, a real existing entity with an assured metaphysical status all of its own.[5]

It is often remarked that although we cannot move a two-dimen-sional figure by a continuous motion in a two-dimensional plane so as to coincide with its mirror image, we can achieve this if we are al-lowed to make use of three dimensions and turn the figure over. In general a reflection in a $(n-1)$-dimensional space can be achieved by a rotation in an n-dimensional space that includes it as a sub-space. If we had access to a fourth dimension, we could move into the looking-glass world. It is natural to wonder whether something of this sort could be achieved in the four-dimensional spacetime of modern physics.

The difficulty is that we do not have any genuinely three-dimensional objects in spacetime physics. What we normally take

---

[4] Graham Nerlich, *The Shape of Space*, Cambridge, 1976, ch.2. We are much indebted to this work for our understanding of the problem and its solution.

[5] See further below, §§9.4–9.5.

to be three-dimensional objects are seen as four-dimensional tubes, extending along a time-like paths with a three-dimensional cross-section. A left hand exists not just for a moment, but over a period of time. If we wanted to turn over a left hand in spacetime, we should have to reverse its temporal direction too. This is just the same as in two dimensions. We can turn over a triangle and make it coincide with its mirror image, but in so doing we have turned it the other way up. If we distinguish its top side from its bottom side, then we realise that the change from being left-handed to being right-handed in two dimensions has been achieve at the cost of a compensating change of direction—from up to down—in the third dimension. So by parity of reasoning[6] a change of handedness in a three-dimensional subspace of spacetime may have to be compensated for by a reversal of the temporal dimension or somehow else. And this indeed is what we find.

Although handedness was important in the history of thought in that it showed the inadequacy of Leibniz's relational account of space, it did not seem to be of any scientific interest. The laws of Newtonian mechanics do not discriminate between left- and right-handed systems. Left- and right-handed screws and spirals and helixes have the same mechanical properties, and angular momentum is conserved either way. We could not tell by any physical experiment involving only Newtonian mechanics whether we were in our ordinary world or in Alice's looking glass world. Only in biology does a difference emerge. Complicated organic chemicals, of the sort found in living organisms, often have molecules that are three-dimensional, and which may differ essentially from their mirror images. Although chemically there is no difference, often only molecules of the one sort are synthesized or utilised by living organisms. And often living organisms show right-handed or left-handed propensities—like runner-beans round their beansticks—But the physicist need not take this as revealing any fundamental preference on the part of nature, but only the accidental fact that the first organism happened to have the one configuration rather than

---

[6] We need to be careful, and remember that spacetime is NOT just like ordinary three-dimensional space: it is topologically markedly different in that the light cone imposes a structure quite unlike anything we have come across in ordinary Euclidean three-dimensional space. Any topological analogies need to be carefully checked before being applied. See above, §3.1

the other. In modern physics, however, the concepts involved have been refined and have been found to play a much more important part than had previously been realised.

Counterparts are mirror images of each other. The operation that takes a figure into its counterpart is reflection. In physics, however, the simple operation of reflection has been replaced by the more subtle one of parity, which can be regarded as an n-dimensional generalisation of reflection. Reflection changes just one dimension, leaving all the others unaltered. If I have a looking glass at the origin and in the plane of the Y- and Z-axes, then the point (x, y, z) will appear to be positioned at $(-x, y, z)$. So it is that if I examine my appearance in a glass, my nose seems to be 12 inches behind the plane of reflection, while my ears, being 14 inches in front of it, seem to be 14 inches behind it, and so my face seems to be looking out of the mirror, while I myself am looking towards it, and thus my front and back are interchanged, which carries with it, in our ordinary way of speaking my left and my right. (That it is only my left and right that are reversed can be seen by describing the parts of my body by means of compass points instead of left and right: if I face North my East arm will be reflected as the East arm of my mirror image, and my West foot will be reflected as the West foot of my mirror image). Although we may have occasion in physics to consider cases in which only one direction has been reversed, it goes against the general *ethos* of physics to treat any one direction differently from any other, and so physicists prefer to deal not with reflections but with changes of "**parity**" in which every co-ordinate has been multiplied by minus one. Thus whereas the reflection of (x, y, z) in the plane of the Y- and Z-axes changes it to $(-x, y, z)$, the parity operation changes it to $(-x, -y, -z)$. So long as we are working in only three dimensions, the difference between a change of parity and a reflection is not great, since a reflection in the plane of the Y- and Z-axes followed by a rotation around the X-axis will achieve the same results as a parity interchange with respect to the origin. Alternatively, we could achieve the same effect by three reflections in the planes of the Y- and Z-axes, of the Z- and X-axes, and the X- and Y-axes, respectively.

Two problems that arose with reflection need also to be raised with parity: can the parity of a particular system change in the ordinary course of events, and do the general laws of physics distinguish at all between one parity and its opposite?

§7.2 Parity

The parity operation P, introduced in the last section, can be generalised to any number of dimensions. It is defined as reflection in the origin, so that a point $(x_1, x_2 \ldots x_n)$ becomes the point $(-x_1, -x_2 \ldots -x_n)$. This is equivalent to successive reflections in n mutually-perpendicular mirrors. The parity operation thus has the same property as the reflection operator that if it is done a second time it brings us back to where we were. Thus for any function $\psi(\mathbf{r})$

$$P\psi(\mathbf{r}) = \psi(-\mathbf{r})$$

operating twice gives

$$P^2\psi(\mathbf{r}) = \psi(\mathbf{r})$$

from which follows its characteristic property that $P = \pm 1$ and so

$$P\psi(\mathbf{r}) = \pm\psi(\mathbf{r})$$

By the 'conservation of parity' we mean that we can carry out the operation P without producing any physically different results. We cannot of course actually reflect a material object in the origin; what we mean is that if we construct two pieces of experimental apparatus, one of which is the exact reflection in the origin of the other, experiments with the one apparatus will give exactly the same results as with the other.

This always seemed obviously true to physicists, and indeed invariance under the parity operation was often used by theoreticians to rule out certain solutions to particular equations which were otherwise formally acceptable. But it had never been tested, until some puzzling results concerning the decay of some elementary particles caused it to be questioned for the first time in the 1950's.[7]

Around 1950 several observations were made of the decay of what were then called heavy mesons, having masses about half that of the proton. One of these concerned the τ-meson, which decays in

---

[7] Allan Franklin in *The Neglect of Experiment*, Cambridge, 1987, recalls earlier parity-violating experiments, but their implications went unrecognised.

a very characteristic way into three pions, and must therefore have negative parity. Another heavy meson, the κ-meson, decays into two pions, and therefore has positive parity.[8] As the accuracy of the measurements improved it became apparent that the τ-mesons and the κ-mesons have the same mass within the experimental uncertainties, and the same charge and spin. This suggested that they are in fact identical, both being kaons, as they are now called, and having two possible modes of decay. The only difficulty about this interpretation of the data was that they have different parity, and this stimulated a thorough examination of the question of parity conservation.

Lee and Yang soon realised that parity conservation had never been rigorously tested and that it is difficult but not impossible to imagine an experiment that would do this. An experiment will only test parity conservation if it embodies a system that is changed intrinsically by the parity operation. An example is an experiment that involves two different sorts of vectors which behave differently under the parity operator. *Polar* vectors, corresponding for example to linear velocities, are altered by the parity operator: *axial* vectors, corresponding to an axis of rotation, are not. Thus if we have a system that decays by emission of two particles, it is represented by two *polar* vectors. Reflection in the origin gives again the same two vectors, with the same angle between them. This may be expressed mathematically by the invariance of $(\mathbf{r}_1 . \mathbf{r}_2)$ under the parity operation: $(\mathbf{r}_1 . \mathbf{r}_2) \rightarrow (-\mathbf{r}_1 . -\mathbf{r}_2) = (\mathbf{r}_1 . \mathbf{r}_2)$. But *axial* vectors behave differently; they are unchanged by the parity operation: thus the angular momentum $(\mathbf{r} \times \mathbf{p}) \rightarrow -\mathbf{r} \times (-\mathbf{p}) = \mathbf{r} \times \mathbf{p}$. But if we have a system that embodies both a polar and an axial vector, the parity operation does not leave it unchanged. Thus cosine of the angle between the vectors

$$\mathbf{r}_1 . (\mathbf{r}_2 \times \mathbf{p}) \rightarrow (-\mathbf{r}_1) . (-\mathbf{r}_2) \times (-\mathbf{p}) = -\mathbf{r}_1 . (\mathbf{r}_2 \times \mathbf{p})$$

that is to say if the angle in the original system is θ, the angle in the system transformed by the parity operator is $180° - \theta$. One such system is the decay of a nucleus spinning with angular momentum I. If parity is conserved we should observe the same number of particles emitted at angles θ and $180° - \theta$ with respect to the direction

---

[8] C.N. Yang, "Present Knowledge about the New Particles", *Reviews of Modern Physics*, **29**, 231. 1957

of the angular momentum vector, as shown in the figure, or to put it another way the angular distribution of emitted particles must be symmetric about 90°. (See Figure 7.2.1.)

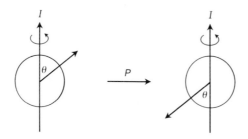

Figure 7.2.1 The effect of the parity operation on particle emission from a nucleus with spin I.

An example of such a process is the β-decay of ^{61}Co. To test parity conservation we thus measure the angular distribution of the β-rays with respect to the direction of the nuclear spins; if it is symmetrical about 90° then parity is conserved. Although so simple conceptually it is technically a difficult experiment, since the nuclei have to be aligned, and this can only be done by very strong magnetic fields at very low temperatures. But, though difficult, it was done by Wu and collaborators in 1957.

The theoretical physicists waited rather anxiously for the result. Pauli, for example, had discarded solutions to the wave equation that do not satisfy parity conservation when developing his theory of the neutrino. If parity turned out not to be conserved he would have to think again. On 17 January 1957 he wrote to Weisskopf. 'I do **not** believe that the Lord is a weak left-hander, and I am ready to bet a very high sum that the experiments will give symmetric results'.[9]

Two days later the experiments were completed, and the distribution was found to be highly asymmetric: 'parity is not conserved', Pauli wrote again to Weisskopf on 27 January 1957: 'Now after the first shock is over I begin to collect myself. Yes, it was

[9] R.Kronig and V.S.Weisskopf, eds., *The Collected Papers of Wolfgang Pauli*, Interscience, 1964, I, xvii.

very dramatic. On Monday 21st at 8.00pm I was supposed to give
a lecture on neutrino theory. At 5.00pm I received three experimental papers. I am shocked, not so much by the fact that the
Lord prefers the left hand, but by the fact that he still appears
to be left-right symmetric when he expresses himself strongly. In
short, the actual problem now seems to be the question why are
strong interactions right and left symmetric.'

There **is** thus a fundamental asymmetry in nature. But is this
all that can be said, or is it but one manifestation of a higher symmetry? It seems that it is. There is another symmetry operator,
that of charge conjugation that converts matter to antimatter and
reverses the directions of all electric currents; and the combination
of the parity operator with that of charge conjugation, PC, appears
to be conserved.

The implications of this can be described by a story. Suppose we
wanted to describe a corkscrew to a friend on a distant star using
only our radio. We can describe it in meticulous detail, giving the
co-ordinates of every part of it to our friend. He could construct an
exact replica and it would be the same in every detail except that we
would never know whether his was a left-handed or a right-handed
corkscrew. The handedness of the corkscrew cannot be described
on the radio by means of ordinary geometry or pre-1957 physics
because we cannot know whether his co-ordinate axes are right- or
left- handed. But now, after the ^{61}Co experiment, we have a way
of describing handedness. We tell him to set up the experiment,
and this will tell him the handedness of his corkscrews. But then
we have an awful thought: suppose he is on an **anti**matter star; if
so the ^{61}Co experiment will come out the other way, and we still
do not know whether his corkscrew is left or right-handed!

This is not the end of the story. In 1964 Fitch and Cronin found a
weak interaction that violates even PC conservation. Once more we
ask if this is part of a yet higher symmetry. Again, it seems that it
is. If we combine PC with time reversal T the combination of time
reversal together with charge conjugation and parity TCP appears
to be conserved. It is both a fundamental result of quantum field
theory and, thus far, consistent with the experimental data.

§7.3   Time Indifference and Time Reversal

The conclusions of the previous section constitute a further argument for supposing that the direction of time, and therefore time itself, is not of fundamental importance in physics. Many thinkers, from Plato to Einstein, have concluded that time is in some way illusory, and that a true understanding of reality would be one not contaminated by the transitory passage of time.[10] Sometimes they are moved by considerations not peculiar to the Special Theory. In our ordinary understanding of time, the present is thought of as special and as fleeting, and the future is thought of as being different, modally and ontologically, from the past, in that whereas the past is unalterable and in some sense undoubtedly exists, the future is open and does not, as yet, exist: in physics, by contrast, as in science generally and the whole of academic knowledge, we seek truths that are not limited to the present time but will be valid at all times for all people in all places. Thus the gossip column of a newspaper is concerned with the people its readers know here and now; but the scientific paper is concerned with observations made at a particular time in a particular place by a particular scientist only in so far as they reveal general truths about the way things happen in given circumstances at any time and any place and observed by anyone. A scientist is, in Plato's words, a spectator of all time, and the universality of his concern requires him to ignore the particularity of the present moment. He is giving us an omnitemporal, and therefore limited, view of reality. There is nothing wrong with that. The mistake lies in forgetting the limitations inherent in the scientific enterprise, and assuming that because the physicist needs to ignore the particularity of the present, therefore everyone else must do so too.

Mathematical physics encourages us to spatialise time. Mathematically time is treated by physicists as a one-dimensional continuum with the order-type of the real numbers, $\Re$. It is easy then to think of it in spatial terms. The pressure to spatialise time, already present in Newtonian mechanics, becomes much stronger in the Special and General Theories, in as much as Minkowski

---

[10] For a collection of scientists' and philosophers' thoughts about the significance of time, see David R. Griffin, ed., *Physics and the Ultimate Significance of Time*, New York, 1986, pp.ix-xv.

spacetime is more highly integrated into a single, and arguably spacelike, whole than the artificially adjoined space-and-time of Newtonian mechanics.[11] The whole theme of spacetime physics in Chapter 1 was to assimilate time and space as far as possible. But the assimilation is not complete, and may make space timelike as much as time spacelike. Although Minkowski spacetime is a four-dimensional manifold, $\Re^4$, its natural topology is very different from a Euclidean four-space. Light cones are fundamental, and signify a profound difference between timelike and spacelike directions. The Lorentz signature is $(+ - - -)$, not $(+ + + +)$, and a pervasive factor of $\sqrt{-1}$ distinguishes timelike from spacelike separations, and velocities, rapidities and pseudo-angles from genuine angles. Spacetime does not really, *pace* Minkowski, abolish the distinction between time and space, and before we objectify it completely, and think of ourselves merely as crawling along one of its worldlines, we should remember other, less geometrical, approaches to the Special Theory, and in particular that of Chapter 3, in which space is more timelike than the reverse.

Quite apart from the geometrical representation of time, the Special Theory seems to argue against there being any real distinction between past, present and future in as much as it depends on our choice of frame of reference how we should assign those terms to distant events. By choosing a suitable frame of reference, events which had been past can be described as future, and *vice versa*. But this, as we saw in Chapter 2, §2.7, is a notational difference only. The events which we can describe as past or future at will are events which are outside the light cone, and which, therefore, we can neither influence nor be influenced by. It makes no difference, as far as the course of events is concerned, how we describe them. The only difference is one of our convenience. By ascribing to them dates depending, in accordance with the Lorentz transformation, on the frame of reference, we can state the laws of electromagnetism in a lucid and coherent form that transcends the choice of any particular frame of reference. This is a very reasonable thing to do, but it does not mean that we are ascribing real properties of pastness or futurity to distant events.

A final difficulty in reconciling the Special Theory with the traditional distinction between past, present and future was that if

---

[11] See above, §§1.1, 1.6.

this distinction were a fundamental worldwide one, it would pick out some frame of reference (perhaps not an inertial one) as a preferred frame. Although certainly electromagnetic, and possibly some other, interactions are Lorentz-invariant, it is not necessary that all should be; preferred frames occur commonly in cosmology and the General Theory, and might well occur elsewhere. Once again it is a misreading of the Special Theory to suppose it is incompatible with our ordinary understanding of time.

Time reversal presents a deeper challenge. If there is no fundamental difference between the direction of time from future to past and from past to future, there can be no real difference between them. And whereas we can accommodate the Special Theory's seeming neglect of the difference as due to the limitations of its own approach, a positive claim that time is fundamentally isotropic, the same in both directions, would, indeed, be inconsistent with our normal understanding of time. Even before the formulation of TCP invariance the Special Theory seemed to be isotropic as regards time, like Newtonian mechanics earlier: and thus there has been a deep debate among physicists on how best to reconcile the purported isotropy of fundamental theory with the evident asymmetries of temporal phenomena.[12]   It is often pointed out that when we consider not the form of the fundamental laws, but their application to physical phenomena, anisotropies appear. Solutions involving advanced potentials are discarded. We are happy to envisage point-sources of electromagnetic radiation, but not point-sinks: although a spherical wavefront contracting inwards to a point is perfectly consistent with Maxwell's equations, we think it too much of a coincidence that a whole set of disconnected and widely separated sources should each have emitted radiation at exactly the right place and time to have set up such a contracting sphere, whereas it is easy to explain an expanding sphere in terms of emission from a single point-source. Our ideal of scientific explanation imposes a constraint on what can count as an explanation, and thus on the direction of explanation, and

---

[12] R.Penrose, "Singularities and Time Asymmetry," in S.W.Hawking and W.Israel, eds., *General Relativity*, Cambridge, 1979, lists seven types of time asymmetry: K^0−meson decay, in quantum mechanics, entropy, retardation of radiation, expansion of the universe, black holes *versus* white holes, and psychological time.

hence also on the direction of time. So too in classical thermodynamics, although the Newtonian laws of impacts between molecules
are isotropic as regards time, the coincidence required for a spontaneous transition from a more to a less disordered state is so great
as to be completely discounted.

This is not the whole of the argument. Besides the argument
from explanation there is an argument from fuzz. Our equations
for natural laws and specifications of boundary conditions are idealised: we give an exact value to the position of an electric charge
or the direction of the electric field at a particular point. But truly
exact values are not available to us. Not only are our measuring
procedures imprecise, but there is outside interference—at the very
least the noise of the rumbling echoes of the Big Bang. In applying
natural laws we have to consider not one absolutely precisely specified boundary condition, but an inexactly specified neighbourhood
of such conditions. And then a marked asymmetry appears. If we
consider a very small disturbance to a point-source emitting radiation, we shall still have a point-source, perhaps in a slightly different
position, quite likely still emitting radiation, though possibly of a
different frequency: whereas, if we were operating with advanced
potentials, and considered a very small disturbance in the sphere
that could radiate inwards towards a point-sink, it would destroy
the prerequisite harmony for the wavefront to contract inwards towards a central point. Not only would it be a great coincidence
for all the disparate sources to radiate at just the right place and
time so as to create an inwardly contracting wavefront, but the
coincidence would be extremely fragile to any inherent variation or
disturbance from outside. The reversibility objection, which has
plagued physicists thinking about thermodynamics, is due to their
idealisation of natural phenomena. To every idealised solution of
Maxwell's equations involving retarded potentials there is indeed
an idealised solution involving advanced potentials: but to every
fuzzy approximation of the one there is not a corresponding fuzzy
approximation to the other. And so, however isotropic the fundamental laws of nature, their application to phenomena has, for
these two reasons, a manifest direction.

Many thinkers have concluded that time in itself is fundamentally isotropic, and its apparent direction is completely accounted
for by the Second Law of Thermodynamics, itself perhaps the out-

come of the Big Bang.[13] Cosmologists cite the expansion rather than contraction of the universe and the predominance of black holes rather than white holes as further instances of temporal asymmetries being due, in the last resort, to boundary conditions rather than the laws of nature themselves. But such a project is open to objection on grounds both of physics and philosophy. The directedness of time not only is given to us in experience as a deep, non-adventitious fact, but is fundamental to the presuppositions of much of our conceptual structure. We think that the future is open and to some extent under our control, whereas the past is unalterable and fixed, so that it is no use crying over spilt milk. We can, however, remember the past, whereas we can only predict the future. We distinguish causal explanations, in terms of antecedent conditions and scientific laws or general regularities, from purposive, or as Aristotle called them teleological, explanations, in terms of some end aimed at. Any view of reality in which the peculiarly temporal characteristics of time played no fundamental part would be one so much at variance with our ordinary experience and understanding that we should have to abandon all our normal ideas about ourselves, our interactions with the world and our knowledge of it.[14]

Nor is it only human ways of thinking that protest against a fundamental isotropy of time. Though many physical phenomena seem to be governed by laws that are isotropic with respect to time, not all are. The decay of K_0-mesons appears to violate conservation of Charge (C) × Parity (P), so that TCP can be preserved only at the cost of the isotropy of Time (T). Of course, we cannot be sure that the question rests here. There always might be some deeper symmetry, analogous to TCP invariance, which accounted for K_0-meson decay without positing a fundamental anisotropy of time. But the very programme of accounting for K_0-meson decay should make us wary. We are, quite largely, imposing rather than discovering time-reversibility: whenever we are faced with a law that is not time-reversible, we posit some further factor to explain it, in much the same way as earlier physicists maintained the isotropy of space in spite of the behaviour of compass needles by postulating

[13] See Hans Reichenbach, *The Direction of Time*, Berkeley, 1956, or more recently Paul Horwich, *Asymmetries in Time*, Cambridge, Mass., 1987.

[14] See above, §3.6, p.112., and below, §9.5, *ad fin*.

a magnetic field. There is nothing wrong with this. But it shows, once again, that much of what the physicist discovers is the consequence of his own stipulations. Just as he may perfectly properly choose to ignore the particularity of the present in his search for omnitemporal generality, so he may always be seeking laws that are isotropic with respect to time as well as space, and not be content until he has found them. But though such a selective search may often be rewarded with success, it does not follow that what is found embraces the whole of reality. Our concept of time, in particular, is connected not only with the physicists' concepts of space, spacetime, and entropy, but with our concepts of action, knowledge and explanation, themselves presupposed by the thinking of physicists. The project of proving time to be fundamentally isotropic is a dubious one, and often, indeed, it seems that physicists having taken much trouble to abstract from our ordinary concept of time a watered-down, isotropic surrogate, then go to great lengths to reintroduce, by appeal to thermodynamics or cosmology, the very directedness they had been at pains to remove.

§7.4 Symmetry Principles and Conservation Laws

In both classical and quantum mechanics there is a deep relation between the properties of a system under various symmetry operations and the conservation of the corresponding physical quantities.[15]

Consider a system, not subject to external forces; whose motions is described by generalised co-ordinates $q_i(t)$ and $\dot{q}_i(t)$. Then the Lagrangian $\mathcal{L}(q_i, \dot{q}_i)$ of the system satisfies the equations

$$\frac{\partial \mathcal{L}}{\partial q_i} - \frac{d}{dt}\left(\frac{\partial \mathcal{L}}{\partial \dot{q}_i}\right) = 0. \qquad (7.4.1)$$

Now suppose that the Lagrangian $\mathcal{L}$ is independent of a particular co-ordinate q_j; such co-ordinates are called cyclic. Then

$$\frac{\partial \mathcal{L}}{\partial q_i} = 0, \qquad (7.4.2)$$

---

[15] H.Goldstein *Classical Mechanics*, Addison-Wesley, 1959, pp.47, 220, 258. L.R.B.Elton *Introductory Nuclear Theory*, Pitman, 1965, p. 43. T.W.B. Kibble *Classical Mechanics*, McGraw-Hill, 1966, p.242. L.L.Landau and E.M. Lifshitz *Mechanics*, tr.J.B.Sykes and J.S.Bell, Pergamon Press, 1960, p.14. L.I. Schiff *Quantum Mechanics*, McGraw-Hill, p.132.

so that conjugate momentum $p_j \equiv \frac{\partial \mathcal{L}}{\partial \dot{q}_j}$ is a constant of the motion. For example, if the system is unaffected by a spatial translation in the x-direction, then $\mathcal{L}$ does not depend on x and so the conjugate momentum

$$p_x = \frac{\partial}{\partial \dot{x}} \left(\frac{1}{2} m \dot{x}^2 \right) = m \dot{x} \qquad (7.4.3)$$

is conserved. Thus spatial symmetry—that is the homogeneity of space—implies the conservation of linear momentum. If the system is unaffected by a rotation through an angle θ, then $p_\theta = \frac{\partial \mathcal{L}}{\partial \theta}$ is a constant. Since the rotational kinetic energy is $\frac{1}{2} m r^2 \dot{\theta}^2$, $p_\theta = m r^2 \dot{\theta}$, which is the angular momentum. Thus rotational symmetry—that is the isotropy of space—implies the conservation of angular momentum about the axis of symmetry.

If the system is unaffected by time, then energy is conserved. This is shown by considering the total time derivative of $\mathcal{L}$ when $\mathcal{L}$ is not an explicit function of time

$$\begin{aligned}
\frac{d\mathcal{L}}{dt} &= \sum_i \frac{\partial \mathcal{L}}{\partial q_i} \frac{dq_i}{dt} + \sum_i \frac{\partial \mathcal{L}}{\partial \dot{q}_i} \frac{d\dot{q}_i}{dt} \\
&= \sum_i \frac{d}{dt} \left(\frac{\partial \mathcal{L}}{\partial \dot{q}_i} \right) \dot{q}_i + \sum_i \frac{\partial \mathcal{L}}{\partial \dot{q}_i} \frac{d\dot{q}_i}{dt} = \sum_i \frac{d}{dt} \left(\dot{q}_i \frac{\partial \mathcal{L}}{\partial \dot{q}_i} \right)
\end{aligned} \qquad (7.4.4)$$

Thus $\left(\sum_i \dot{q}_i \frac{\partial \mathcal{L}}{\partial \dot{q}_i} - \mathcal{L} \right)$ is constant.

If now we consider only conservative systems for which the potential energy V is independent of the velocities $\dot{q}_i$. So that

$$\frac{\partial \mathcal{L}}{\partial \dot{q}_i} = \frac{\partial T}{\partial \dot{q}_i}. \qquad (7.4.5)$$

Since T is a homogeneous quadratic function of the $\dot{q}_i$, by Euler's theorem

$$\sum_i \dot{q}_i \frac{\partial T}{\partial \dot{q}_i} = 2T. \qquad (7.4.6)$$

Thus $2T - \mathcal{L} = 2T - (T - V) = T + V$ is constant, and this is the total energy of the system. Thus symmetry under temporal transition—that is the homogeneity of time—implies the conservation of energy.

These three examples are particular cases of a more general theorem about the condition that must be satisfied for a quantity to be conserved. The proof of this theorem requires the corresponding transformation to be continuous, so in classical mechanics it is not possible to obtain the conservation laws such as that of parity associated with discontinuous or discrete formations.

An important example of a discrete transformation is reflection in the origin, represented in Cartesian co-ordinates as $(x, y, z) \rightarrow (-x, -y, -z)$ and in spherical polar co-ordinates as $\mathbf{r} \rightarrow -\mathbf{r}$. This transformation corresponds to the parity operator and to show parity is conserved we have to use a quantum-mechanical argument.

We consider a system described by the wavefunction $\psi(\mathbf{r})$. The effect of the parity operator P is given by $P\psi(\mathbf{r}) = \psi(-\mathbf{r})$ where $P = \pm 1$. Thus all systems are characterized by either positive or negative parity. If the parity operation is applied to the Schrödinger wave-equation, then provided the Hamiltonian is unaffected we obtain

$$H\psi(\mathbf{r}) = E\psi(\mathbf{r}), \qquad (7.4.7)$$

and

$$H\psi(-\mathbf{r}) = E\psi(-\mathbf{r}), \qquad (7.4.8)$$

so that the wavefunctions $\psi(\mathbf{r})$ and $\psi(-\mathbf{r})$ are both associated with the same eigenvalue E. If the state is non-degenerate, the wavefunctions must be proportional to each other, so that

$$\psi(-\mathbf{r}) = K\psi(\mathbf{r}). \qquad (7.4.9)$$

Thus by the same argument as before, $K = \pm 1$.

However it can be written

$$P\psi(\mathbf{r}) = K\psi(\mathbf{r}), \qquad (7.4.10)$$

which is an eigenvalue equation for the parity operator. Thus the parity must be a constant of the motion, having the value $+1$ or -1.

More generally, it is possible to show quantum-mechanically that the condition that an observable is a constant of the motion is that its operator commutes with the Hamiltonian of the system, providing that operator is not explicitly time-dependent. Thus the expectation value of a variable r with operator R in a system with wavefunction ψ is

$$r = <\psi \mid R \mid \psi>. \qquad (7.4.11)$$

So that

$$\dot{r} = < \dot{\psi} \mid R \mid \psi > + < \psi \mid \dot{R} \mid \psi > + < \psi \mid R \mid \dot{\psi} > . \quad (7.4.12)$$

The wavefunction satisfies the time-dependent Schrödinger equation

$$H \mid \psi > = i\hbar \mid \dot{\psi} > . \quad (7.4.13)$$

So that

$$\dot{r} = \frac{i}{\hbar} < \psi \mid HR - RH \mid \psi > + < \psi \mid \dot{R} \mid \psi > . \quad (7.4.14)$$

Thus r is a constant of the motion provided R commutes with the Hamiltonian and is not explicitly time-dependent. This condition may be used to derive quantum-mechanically the conservation laws already obtained classically.

§7.5 Sameness and Difference

The conservation of parity is a moral tale. One moral is how strongly physicists are wedded to certain basic principles. Pauli, like Oersted before him, had a clear idea of what ought to be the case. He did not passively wait on the experimentalist to tell him what he should believe, but in advance of the empirical evidence had a firm expectation of what its outcome was going to be, and was prepared to bet a very high sum that the experiments would give symmetric results. It is an old tale, but the fact that physicists continue to stick their necks out in the knowledge that when their predecessors have done so they have sometimes been proved wrong, witnesses to the strong *a priori* strand in physical thinking.

A second moral is the extent to which physicists are prepared to save their principles in the face of adverse experimental evidence by invoking further considerations which will make the principle at risk still true at a deeper level. In the previous chapter[16] we saw how Ampère's law was massaged to get it into the right shape to be Lorentz covariant. When the conservation of parity was found not to fit all the experimental facts, it too was "massaged", by being enveloped in a deeper symmetry, not just parity, P, but parity combined with charge conjugation, PC. And when later PC was in

[16] §6.7

turn found not always to hold, it too was saved from crude falsification by being subsumed under the even more general principle of the conservation of time reversal together with charge conjugation and parity, TCP. If the experimental evidence fails to fit a cherished principle, we do not just abandon the principle, but reckon that we have not got hold of it quite right, and try to reformulate it in a more profound way which will accommodate the adverse evidence. We tend to go on believing that there is some important similarity in nature, and attempt to refine the specification of it. The two different systems are really similar, after all, we reckon, but the relevant condition is not that they should be transformed into each other by a parity operator; it is rather that they should be transformed into each other by a combination of operators of which parity is one, so that, if in some interaction the others are not conserved, then parity will not be conserved either.

There is a third moral to be drawn, a more moralistic one. Physicists doom themselves to perpetual unhappiness, because they are always seeking deep similarities, and then feel worried by the absence of differences they think should still be there. Leibniz, and later Kant in some of his phases, felt that it was a defect of Newtonian space that it could not distinguish in purely geometrical terms between a right hand and a left hand. There seemed to be an inherent distinction without a difference, which called in question the real existence of space as a substantial entity at all. In the same spirit physicists before 1957 were worried at the thought of a left-handed looking-glass world which was physically indistinguishable from ours but yet undeniably different from it. Once human beings, or human artefacts, such as corkscrews, or biological systems such as runner-beans, were introduced, it would be possible to tell worlds of different parity apart. But there was no difference, so far as *physics* was concerned, and two rational agents who could communicate only by radio would never be able to determine whether they were of the same, or of opposite, parity. Such physicists should have been glad to hear of Wu's discovery of ^{61}Co's having a beta decay that afford a means of distinguishing systems of opposite parity. Instead they compound their worries, and have nightmares about meeting extra-terrestrial beings composed of anti-matter; or, at least within the realms of science fiction, of encountering beings who move backwards in time.

Does this show that physicists are in need of psychiatric treatment for self-induced neuroses? No. It shows rather the extreme difficulty they, and all of us, have in dealing with the concepts of sameness and difference. In the first place we forget that sameness and difference are triadic relations, and two things can be the same as each other in one respect, and different from each other in another respect. The physicist is always seeking deeper and deeper similarities, but these if they are to exist at all, must exist between things which are also in some other respects different. We have deep intimations that space and time are homogeneous and isotropic, and that therefore it should not make any difference if we move things around in space, or compare different experiments carried on at different times. But although it should not *make* any difference, there must *be* some difference, namely that difference which should not make any (physically relevant) difference. It is thus inherent in the idea of physics that there should be some differences which we can recognise but which do not make any physically relevant difference.

It is easier to come to terms with this, if we consider the continuous transformations, such as spatial displacement. In saying that space is homogeneous, we are saying that it is all the same as far as physics is concerned whether an experiment is undertaken in Oxford or in Cambridge. The only factors that can affect the result must be some difference in conditions other than the bare fact that one was in Oxford and the other in Cambridge. But although we are moderately intrigued by the thought of extra-galactic beings who are very far away, we do not find them philosophically perplexing because they are *only* a long way off, and if we had sufficient time we could in principle overcome the difference between them and us. A difference of parity cannot be similarly overcome. Because it is a discrete difference, we cannot overcome it bit by bit. It constitutes a difference not of degree but of kind, which therefore seems much more different. But in the case of parity, it appears also to be much less of a difference, since parity apart there is no difference of any discriminable sort between the two systems. A parity difference is the least of all possible differences, and yet one that can never be overcome by anything we know how to do. Hence the tantalizing nature of the looking- glass world.

We can see why parity is puzzling. But why should it be so pervasive in the physical world? The answer lies in the Theory of Groups. Almost all groups have a sub-group just half as large as they are themselves. In particular, once we have a multi-dimensional space, we can interchange two of the dimensions and so obtain, in effect, left- and right-handed transformations which will resemble each other in every respect save this one. Thus once we have symmetries, we have maximal symmetries, that hold between things which are as like each other as possible without being identical, and these being discrete are unsurmountable, and seem profoundly mysterious in consequence.

Chapter 8
Retrospect and Review

§8.1 Refutation or Refinement

Einstein, it is popularly believed, completely overthrew Newton:

> Nature and Nature's laws lay hid in night:
> God said, Let Newton be! and all was light.
> It did not last: the Devil howling 'Ho!
> Let Einstein be!' restored the status quo.

That is a mistake. It would be nearer the mark to say that, if we approach the theories from the standpoint of the Equivalence Principle,[1] Newton's physics was refined, rather than refuted, by Einstein's Special Theory. The Special Theory needs to be seen both as a development and as a replacement of Newtonian mechanics, which in its time was both a development of, and was seen as a repudiation of, the mechanics which preceded it. In this development the Special Theory and Newtonian mechanics occupy a similar position and are in many important respects alike, even though there are other perspectives —the integrationist approach of Chapter 1, for example— from which their differences are more apparent.

In this chapter we consider first (in §§8.2-8.5) the underlying respects in which the Special Theory and Newtonian mechanics resemble each other and differ from the theories that preceded and came after them, and then ways in which the Special Theory seems to depart from classical orthodoxy, but can be shown to give a different though equally acceptable account of the nature of things.

§8.2 The Lorentz and Galilean Transformations

This section presupposes §2.6

The similarities between the Special Theory and Newtonian mechanics can best be approached by comparing their characteristic transformations. The characteristic transformation of the Special Theory is the Lorentz transformation:

---

[1] As in ch.2.

$$x' = \gamma(x - vt)$$
$$y' = y$$
$$z' = z \tag{8.2.1}$$
$$t' = \gamma(-x\frac{v}{c^2} + t),$$

or, in matrix notation,

$$\begin{pmatrix} \gamma & 0 & 0 & -\gamma v \\ 0 & 1 & 0 & 0 \\ 0 & 0 & 1 & 0 \\ -\gamma\frac{v}{c^2} & 0 & 0 & \gamma \end{pmatrix} \begin{pmatrix} x \\ y \\ z \\ t \end{pmatrix} = \begin{pmatrix} x' \\ y' \\ z' \\ t' \end{pmatrix} \tag{8.2.2}$$

The characteristic transformation of Newtonian mechanics is the Galilean transformation:

$$x' = x - vt$$
$$y' = y$$
$$z' = z \tag{8.2.3}$$
$$t' = t.$$

or, in matrix form,

$$\begin{pmatrix} 1 & 0 & 0 & -v \\ 0 & 1 & 0 & 0 \\ 0 & 0 & 1 & 0 \\ 0 & 0 & 0 & 1 \end{pmatrix} \begin{pmatrix} x \\ y \\ z \\ t \end{pmatrix} = \begin{pmatrix} x' \\ y' \\ z' \\ t' \end{pmatrix}. \tag{8.2.4}$$

It is a special case of the Lorentz transformation, where $c = \infty$. If we want to stress the similarities between the two transformations, we can say that the Lorentz transformation is a more general version and that it approximates to the Galilean transformation for ordinary velocities such as we normally come across in terrestrial phenomena: from this perspective the Special Theory has refined, rather than refuted, Newtonian mechanics. If, however, we want to emphasize the contrast, we can say that whereas Newtonian mechanics assumes implicitly that $c = \infty$, the Special Theory assigns it a definite finite value, 186,000 miles per second, and therefore holds that time depends on position as much as position on time, thus avoiding the asymmetry of the Galilean transformation and the complete separation of time from space: from this perspective the Special Theory has refuted, rather than refined, Newtonian mechanics.

§8.3 The Equivalence Principle

This section presupposes
§2.3 and develops §3.6

The Lorentz and the Galilean transformation both depend on v, the uniform velocity between two inertial frames of reference, and nothing else, and are both intended to be such that natural laws shall be covariant under them. That is to say, they both are means whereby we can regard different frames of reference, moving at uniform velocities with respect to one another, as being essentially the same, so far as certain physical theories are concerned. It is in regard to this Equivalence Principle that the Special Theory and Newtonian mechanics most importantly resemble each other and differ from the physics that preceded and followed them.

There could be other equivalence principles besides Galileo's, either narrower or wider. It would be very natural to suppose, as it was supposed by Galileo's predecessors, that any velocity, even a uniform velocity, did make a difference to what we observe. Indeed, it would be quite natural to think that position, orientation and date are in themselves physically significant, as Aristotle thought.[2] According to Aristotle each thing had its own place, and, if not in it already, would seek to get there; thus stones belonged to the nether regions and had a natural tendency to fall towards them, while fire belonged to the uppermost region and had a natural tendency to rise. In the same way later thinkers placed Jerusalem at the centre of the universe, reckoned that East was the best direction and expected the Big Crunch to come in exactly 1000 AD.

In the later Middle Ages the works of Aristotle became known in Western Europe, and stimulated intense debate among scholastic philosophers and theologians. Aristotle provided the most sophisticated world-view then available, and this was used, particularly by Aquinas, to build a synthesis of secular and divine knowledge. Difficulties arose, however, when it was realised that some of Aristotle's ideas were contrary to the Christian faith, and in their commentaries the medieval philosophers did not hesitate to differ from him for theological reasons. In 1277 the bishop of Paris, Etienne Tempier, condemned 219 philosophical propositions as inconsistent

---

[2] We are not being fair to Aristotle in applying his name to the proto-physics that comes first in our schematic account.

with the Christian belief that God created all things out of nothing. In particular, he condemned the claim that God could not move the world with a rectilinear motion because a vacuum would be left behind, and thus "Aristotle's argument for the necessary existence of a single centre and circumference, on which he had founded his belief in a unique world, was plainly subverted."[3] Jean Buridan extended the argument against any particular places or particular directions having special significance: "God, when He created the world, moved each of the celestial orbs as He pleased, and in moving them He impressed on them impetuses which moved them without His having to move them any more except by the method of general influence whereby He concurs as co-agent in all things which take place."[4] In this Buridan not only argued for the homogeneity and isotropy of space on grounds of divine omnipotence but adumbrated Newton's first law of motion in a qualitative way.

Galileo extended the range of the Equivalence Principle by holding also that systems that differed only in respect of being in uniform motion with respect to one another were the same so far as the laws of physics were concerned, and both Newtonian mechanics and the Special Theory are theories for which this principle of equivalence holds good. It would be quite natural to consider extending the equivalence principle further to include uniformly accelerating systems; and further again, to include those for which the change of acceleration was constant; and ... and; for every derivative of position with respect to time, we can stipulate that systems differing only in that respect should be equivalent so far as the laws of physics are concerned, and that therefore the laws of physics should be covariant under the appropriate transformation. This essentially is what Einstein did in formulating his Principle of Covariance which is the cornerstone of the General Theory, which

---

[3] E.Grant, *The Cambridge History of Later Medieval Philosophy*, Cambridge, 1982, p.537

[4] Quoted by M.Clagett, *The Science of Mechanics in the Middle Ages*, Madison, University of Wisconsin Press, 1959. Buridan was followed by his pupil Nicole Oresme, who likewise denied the unique direction of motion: "If God in His infinite power created a portion of earth and set it in the heavens, this earth would have no tendency whatever to be moved towards the centre of our world;" see N.Oresme, *Le livre du Ciel et du Monde*, Madison, University of Wisconsin Press, 1968.

we can view as the extension of the Equivalence Principle to its ultimate limit, in which not only the first but every derivative of position with respect to time makes no difference to the form of physical laws.

Table 8.3.1   **The Scope of the Equivalence Principle**

Do physical laws take a different form in systems which differ in respect of ...?				
	Change of Origin	Change of Axes	Uniform Velocity	Acceleration
Aristotelian Physics	yes	yes	yes	yes
Mediæval Mechanics	no	no	yes	yes
Newtonian Mechanics	no	no	no	yes
The Special Theory	no	no	no	yes
The General Theory	no ?	no ?	no ?	no ?

In this table we see what changes make a difference to the physical phenomena, and what changes leave them essentially unaltered. Galileo argued that just as a change of position made no difference to the way a physical system behaved, so it did not matter whether it was moving uniformly in a straight line or whether it was at rest. In this crucial respect the Special Theory is, along with Newtonian mechanics, a Galilean theory, and as such to be sharply distinguished from Aristotelian and medieval physics before it and from the General Theory after it.

The entries in the last line of table 8.3.1 are problematic. Although Einstein held it to be the aim of the General Theory to extend the Equivalence Principle so as to include accelerating frames of reference represented by co-ordinate systems with curvilinear co-ordinates, this characterization of the General Theory is

inadequate.[5] It is beyond the scope of this book to give a proper
consideration of the General Theory, and we are concerned to view
it only from the standpoint of the Special Theory. But from this
standpoint we can remark that the wide ambitions of the General
Theory compromise its success in achieving more limited aims. The
General Theory allows a far wider range of co-ordinate systems—
languages. Any co-ordinate system will do, provided it is continuous and distinguishes timelike, lightlike and spacelike trajectories,
but the price for such extreme generality is that we have, in effect,
to take cognizance of the languages involved in translating from
one to the other—the law has to contain terms which depend on
the affine connection and metric tensor. In the Special Theory,
by contrast, there is a much sharper distinction between what depends on the particular co-ordinate system being used and what
does not. Provided we limit ourselves to rectilinear, orthogonal
co-ordinate systems, then although the actual co-ordinates used to
refer to particular events will—of course—depend on the particular co-ordinate system adopted, the magnitude of their spacetime
separation, and the form of the laws of Newtonian mechanics and
Electromagnetism will not. This constitutes a much more cogent
constraint upon admissible laws, and so tells us much more about
the form, and hence also about the content, of the laws themselves.
Einstein's Principle of General Covariance tells us less, because in
order to accommodate all continuous one-one mappings, it has to
be so flexible that almost any putative law could be made to satisfy
it.[6] If very curvilinear co-ordinates are to be admitted, we must
take account of curvature in expressing physical laws, and then cannot argue, as we did in deriving the Lorentz transformation, that
the transformation must depend only on v, or that the electromagnetic potentials may "potentially depend only on the position
and the velocity, and not, for instance, on the acceleration".[7] It
is the strength, but also a weakness, of the General Theory that
it is a theory of gravitation, and therefore is concerned with acceleration and its further derivatives, and so comes to be able to
distinguish, in terms of these derivatives as they appear in the

[5] See R.Torretti, *Relativity and Geometry*, Oxford, 1983, ch.5., pp.135-137.

[6] See above, §2.3, p.36.

[7] See above, §6.7, p.204.

affine connection and the metric tensor, every point of spacetime from every other. We lose the homogeneity of spacetime. Whereas electromagnetic phenomena are the same in Cambridge as in Oxford, the gravitational geodesics are inevitably different, at least in direction, and active transformations[8] are invalid, since the gravitational background is typically different. The Special Theory, by distinguishing between the spacetime background and the physical laws that are effective, is able to require physical laws to be much more definite in shape and content—the homogeneity of space and time was one of the highroads to the Lorentz transformation—its limited ambitions are also its source of strength.

§8.4   The Causal Inefficacy of Space and Time

This section develops the argument of §5.2

The Equivalence Principle gives rise to two theses about space and time, which we may call the Causal Inefficacy of Space and Time and the Causal Opaqueness of Space and Time

The doctrine of the causal inefficacy of space and time holds that spatial position and orientation and temporal position are *per se* irrelevant to physics. There is no preferred position or direction or date. That a physical phenomenon occurs where and when it does is due not to its absolute spatiotemporal position, but to its spatiotemporal relation to other non-spatiotemporal factors, such as a pulse of electromagnetic radiation or a magnetic field. Many arguments can be adduced in favour of the causal inefficacy of space and time,[9] but there are difficulties in seeing exactly how far it should extend.

The first difficulty concerns derivatives. Although we are fairly easily persuaded that position, orientation and date should not by themselves make any difference to physical phenomena, we do not find it obvious that velocity should not. It is not an absolute conceptual truth that physical phenomena in frames of reference moving with a uniform velocity with respect to one another should be the same: witness the fact that they are not the same if the frames of reference are not moving with uniform velocity but are

---

[8] See above, §2.4.

[9] See below, §§10.4, 10.5.

accelerated or rotating.[10] Clearly, then, it is not part of the concept
of time and space that they are altogether causally inert, so that the
physicist can never have occasion to believe that they are real, but,
rather, it is a conceptual truth about the laws of physics that they
should be covariant under change of origin or orientation and, for
Newtonian mechanics and electromagnetism, under change from
one frame of reference to another moving with uniform velocity
with respect to it.

The second difficulty in the doctrine of the causal inefficacy of
space and time concerns scale. If position, direction and date are
causally irrelevant, why not distance, angle and duration? We take
it for granted that an egg can be boiled today in the same way as it
could yesterday, but that boiling it for six minutes produces a very
different result from boiling it for only three. But why should an
omnipotent God not be able to hard-boil an egg in three minutes
or soft-boil it in six? Why should not the laws of physics remain
the same under change of scale as well as under change of origin?
Galileo pointed out that if an elephant were doubled in size in
each dimension, its mass would be multiplied by 8, while the cross-
section of its bones would be multiplied by only 4, and would be
unable to carry the increased weight. In physics there are certain
fundamental constants, such as Planck's constant, or the speed
of light, which preclude the possibility of a general scaling up or
scaling down.

Scale-indifference plays an important part in differentiating the
parts played by geometry and physics in explaining the course of
events. Although originally geometry, as the Greek word suggests,
was concerned with surveying land, it progressed to being the study
of shapes irrespective of size, while distance and duration have always been regarded as physically significant magnitudes. Geometry
was to be as accommodating as it could be, putting as few constraints as possible upon the way we referred to and characterized
positions and figures in space, while leaving to physics the task not

---

[10] This was the point Newton was making with his bucket experiment, and
which Clarke subsequently made in his controversy with Leibniz in rebuttal
of verificationist arguments against the reality of time and space. See Isaac
Newton, *Principia*, Scholium to Definition VIII, reprinted in H.G.Alexand-
er, *The Leibniz-Clarke Correspondence*, Manchester, 1956, pp.157-160. See
also above, §1.1, fn.2.

just of describing, but of explaining why phenomena were as they actually were.

If this difference of *rôle* is accepted—and it is a big 'if'—geometry needs to be subject to more symmetries than physics. As characterized by geometry, space is more homogeneous than we have hitherto specified, being scale-indifferent as well as origin- and orientation-indifferent: whereas physics, if it is to be mathematical physics, explaining events economically in mathematical terms, needs to regard at least magnitudes, if not absolute spatiotemporal locations, as relevant.[11] Such a distinction is characteristic of Newtonian mechanics. But it is not conceptually necessary to insist on it. On the one hand, non-Euclidean geometries of constant curvature, which are date- and direction-indifferent, are conceptually coherent, and on the other, it is not obviously a necessary truth that there should be fundamental physical constants such as the mass of the electron. In the Special Theory the distinction is slightly eroded in that Minkowski spacetime has a very special topology, which in fact also determines the metric.[12]　In the General Theory the distinction is entirely abolished, and in geometrodynamics we seek to explain the whole course of events by reference to the geometry of spacetime.

§8.5　The Causal Opaqueness of Space and Time

Equivalence Principles restrict, as well as facilitate, causal explanation. Once we accept the minimal equivalence principle that position, direction and date are in themselves matters of scientific indifference, we abjure the hope of explaining scientifically why phenomena happen where and when they do. The laws of nature—that constitute genuine scientific explanations—are covariant under change of origin and reorientation of axes: what determines the where and when of particular phenomena cannot, therefore, be the laws of nature themselves, but must be the initial (or, more generally, the boundary) conditions, which in turn can be explained, granted the laws of nature, only by other initial conditions. There is a certain brute inexplicability about initial conditions which is the other side of the coin of the physicists' search for generality.

---

[11] See below, §10.5.

[12] See above, §§3.2–3.4

The laws of nature are general, and therefore cannot by themselves explain particular phenomena in all their particularity. To explain why something happened there and then we have to invoke some other particularity, not itself explicable in general terms alone.

There are thus limits to what a scientific theory that accepts some Equivalence Principle can explain, and concomitantly to what it regards as calling for explanation. Aristotelian proto-physics is very wide in what it purports to explain: the reason why the air is above the earth is that is its proper place; the reason why flames go upwards is that they are trying to get to their proper place. But if the laws of nature operate the same everywhere and everywhen, we can no longer hope to answer in scientific terms questions about the where and the when of natural phenomena. Why did the Big Bang happen about 15,000,000,000 years ago and not half a giga-year sooner or later? Leibniz held that it was a meaningless question,[13] but like many awkward questions, it can intelligibly be asked: the only difficulty is that we cannot answer it.

Galileo's Equivalence Principle restricts further the range of what can be, and needs to be, be explained. Whereas for the minimal equivalence principle we cannot explain why things are where they are, and do not need to explain their remaining where they are, but only their moving, for Galileo's Principle uniform motion in a straight line neither calls for, nor admits of, any explanation. Accelerations still have to be explained in terms of forces, but motion with uniform velocity is like rest: it calls for no explanation; it is just a matter of happenstance, due to some initial condition, perhaps laid down by divine *fiat* at the dawn of creation.

The later Einstein altered again the agenda of physics. He asked new questions—why is the inertial mass of a body the same as its gravitational mass?—and unasked some old ones. The fact that the planets do not move with uniform velocity is no longer something that calls for special explanation in the General Theory. It is part of the fabric of the universe that they should follow a geodesic path in a curved spacetime. Being accelerated is nothing out of the ordinary; it is only natural: what would call for explanation

---

[13] See further below, §9.3, pp.257-258.

would be a sudden[14] change of acceleration. A rocket in free fall is just as unremarkable the General theory as a stone sliding with uniform velocity on a sheet of ice is in Newtonian mechanics or in the Special Theory, or a mountain remaining still in pre-Galilean physics. What would call for explanation would be the rocket's moving out of its natural orbit: such an unnatural acceleration would be attributed to the propulsion of its motors and the speedy discharge of its fuel.

As Principles of Equivalence are extended, it is arguable that the explanations offered by the corresponding theories become deeper, but it is incontestable that the range of opaque initial conditions is increased. In Newtonian mechanics we have to specify the velocities of each particle as well as the position, whereas in pre-Galilean physics it would have been thought that to specify the position would be enough. There are conceptual advantages in having a greater number of initial conditions. It gives a greater degree of freedom, so to speak, and enables us to identify a physical system— which requires the specification of all the relevant positions at some particular time—without thereby committing ourselves to its future evolution. There are also some virtues in having the fundamental equations of change, Newton's Second Law and the wave equation, as second-order differential equations. But there is a price to be paid in the increased range of conditions that have to be taken as brute facts, and as we go beyond Galileo's equivalence principle the price becomes very great.

From the standpoint of Equivalence Principles the Special Theory is very like Newtonian mechanics, both occupying a similar position between medieval mechanics on the one hand and the General Theory on the other. And it is arguable that Galileo's Equivalence Principle, which characterizes them, is as wide as it can be without seriously compromising its explanatory power.

---

[14] That is, non-geodesic. The General Theory can accommodate changes of acceleration, as for instance in a comet orbiting the sun, but only those represented by a smooth geodesic trajectory in spacetime. A kink in a trajectory would call for explanation.

§8.6 Newton's Doctrine of Space and Time

This section picks up the argument of §§1.1 and 1.6

The main difference between the Special Theory and Newtonian mechanics is that in the Special Theory space and time are integrated into a unified spacetime, whereas in Newtonian mechanics they are separate and distinct. In the Special Theory we are dealing with a four-dimensional spacetime, $\Re^4$, with Lorentz signature $(+---)$, whereas in Newtonian mechanics we are dealing with two separate entities, a one-dimensional time and a three-dimensional space, and although we can take the Cartesian product of them, $\Re^1 \times \Re^3$, they remain essentially distinct. Although, as we argued in Chapter 1, the integration of space with time into a unified spacetime is a theoretical advantage for the Special Theory, intuitively they are distinct, and any attempt to make out that time is merely a dimension on a par with those of space is inherently implausible as well as involving us in many difficulties. And in fact, the Special Theory acknowledges some difference between time and space in as much as Minkowski spacetime has a Lorentz signature.

Newton, as we have seen,[15] thought of space as the Divine *sensorium*:

> Does it not appear from phenomena that there is a Being incorporeal, living, intelligent, omnipresent, who in infinite space, as it were His sensory, sees the things themselves intimately, and thoroughly perceives them, and comprehends them wholly by their immediate presence to Himself: of which things the images only carried through the organs of our sense into our little sensoriums, are there seen and beheld by that which in us perceives and thinks.

His space is a Euclidean three-space extending to infinity in all directions. It is highly homogeneous, being indifferent as to origin, direction and scale. It is, arguably, more amorphous, more featureless than Minkowski spacetime or the variably curved spacetimes of the General Theory, and so can be said to put minimal constraints on the activities of an omnipotent God.

Newton's concept of time is similarly appropriate to the Divine consciousness. "Absolute ... time ...," he said, "flows evenly

---

[15] §1.6.

and equably from its own nature and independent of anything external."[16] It is a definition it is easy to pick holes in, but difficult to reject altogether. "Evenly with respect to what?" we may well ask. And yet we have some idea of what Newton was trying to say. It makes sense to wonder whether time might not be speeding up, like the τ-time of some cosmologists,[17] and to maintain that in fact it is not. In saying that time flows evenly and equably, we are sticking our necks out: we expect there to be a unique basic rhythm of the universe,[18] the same for all phenomena in all places, but could discover that there was not, or might be led to abandon the claim by other considerations not thus far envisaged. Newton's definition gives us some guidance, both on what to expect and on what not to expect. And indeed we find that our different means of measuring time—by mechanical clocks, quartz crystal clocks and atomic clocks—correlate with one another to high accuracy; and so we believe that we are finding increasingly accurate means of measuring a real physical quantity; that is, that "relative, apparent and common time"—the time established by our actual measurements—is asymptotically approaching absolute, true and mathematical time. More immediately to our purpose, Newton's last phrase, 'independent of anything external' has some, negative, content. It excludes the last term of the Lorentz transformation

$$t' = \gamma(t - vx/c^2);$$

t' cannot depend on x or v, but must be independent of everything except t itself.

---

[16] I.Newton, *Principia*, Scholium to Definition VIII, in H.G.Alexander, ed., *The Leibniz-Clarke Correspondence*, Manchester, 1956, p.152.

[17] See E.A.Milne, *Proceedings of the Royal Society*, series A, **158**, 1937, 324-348; and *Kinematic Relativity*, Oxford, 1948, esp.§ 30, p.36; P.A.M.Dirac, *Proceedings of the Royal Society*, series A, **165**, 1938, pp.199-208; E.Teller, *Physical Review*, LXXIII, 1948, pp.801-802; M.Johnson, *Time and the Universe for Scientific Conscience*, Cambridge, 1952; and D.H.Wilkinson, *Philosophical Magazine*, III, 1958, pp.582ff.; and G.J.Whitrow, *The Natural Philosophy of Time*, 1st ed. Edinburgh, 1961, ch.5., pp.223-267; 2nd ed., Oxford, 1980, ch.6, pp.270-320, and §7.7, pp.361-364.

[18] The phrase comes from G.J.Whitrow, *The Natural Philosophy of Time*, 1st ed. Edinburgh, 1961, ch.1, §10, p.46; 2nd ed., Oxford, 1980, p.44.

The independence of Newtonian time from Newtonian space is the most important feature that differentiates Newtonian mechanics from the Special Theory. Whereas simultaneity is relative to the frame of reference in the Special Theory, in Newtonian mechanics it is absolute. If, as in the last equation of the Galilean transformation, the time co-ordinate does not depend at all on the relative velocity or the spatial co-ordinates, then we can determine classes of simultaneous events all over the universe which do not depend on either the position of the events or the relative velocity of one frame as compared with another. Simultaneity is thus independent of the frame of reference in Newtonian mechanics, and is in that sense absolute; from this it follows that the date assigned to any event is independent of both its position and the frame of reference we are using, and hence also the duration of the temporal interval between two events. Absolute simultaneity is often held to be incompatible with the Special Theory. But this, as we have seen,[19] is a misapprehension of the proper range of the Equivalence Principle. No inconsistency results if we add a canon of absolute simultaneity to the Special Theory: rather, the effect is to pick out a preferred frame of reference, and thus a canon of absolute rest. In this respect, somewhat surprisingly, Newtonian mechanics is less absolute than the Special Theory. The Newtonian canon of absolute simultaneity does not carry with it any canon of absolute rest, but is compatible with any inertial frame of reference. Although Newton himself was aware that his mechanics gave him no criterion for determining absolute rest, and that he must rely on God's omniscience, or else on some arbitrary choice, to give content to the notion of rest, and hence to that of an object's being in the same place at different times, standard expositions of Newtonian mechanics have not, until recently, taken proper account of this difficulty. Neo-Newtonian space is messier than Newton made out.[20] To that extent, the argument from simplicity in favour of Newtonian mechanics rather than the Special Theory has to be qualified. In general, however, the argument from simplicity tells

---

[19]   §3.9, p.118.

[20]   For further accounts, see L.Sklar, *Space, Time, and Spacetime*, University of California Press, 1974, ch.III, §§D 3,4, pp.202ff.; H.Stein, "Newtonian Space-Time", *Texas Quarterly*, **10**, 1976; J.Earman, "Who is Afraid of Absolute Space?", *Australasian Journal of Philosophy*, **40**, 1970.

against the argument from integration adduced in Chapter 1. It is
the experimental evidence in favour of the Special Theory together
with its unifying power with regard to electricity and magnetism
that tips the balance in favour of the Special Theory.

The argument of this section is developed further in §9.5

§8.7 Newton's Theology; and Einstein's

The metaphysical and theological suggestions of a physical theory
are only tentative. Seen in a different light, the same theory may
suggest a different metaphysics. And in any case other physical
theories may have different metaphysical implications, or may alter
the implications of the theory in question. Any conclusions we draw
from any particular scientific theory may well be wrong. But it is
only by attempting to follow out all its implications that we obtain
a full understanding of a theory, and it is only by running the risk
of being wrong that we can ever hope to be right.

Newton's theology and physics were all of a piece. Believing in
an omniscient God, it was natural to conceive of all points of space
being alike immediately presence to His consciousness; and only
by the arbitrary *fiat* of an omnipotent God could he explain the
initial configuration of the atoms at the dawn of creation. The
Newtonian scientist naturally takes the God's-eye view of nature,
and for him such a perspective is intuitively accessible. Newton's
theism was simple, almost *simpliste*, so much so that he feared he
would be thought heretical, and indeed was widely suspected of
unitarianism. But whatever its theological inadequacies, Newton's
theism was intelligible, and fitted in with his physics.

The metaphysical implications of the Special Theory are much
less clear. Einstein himself, though deeply religious, with a "humble attitude of mind towards [the] grandeur of reason incarnate in
existence", and maintaining that the "highest principles for our aspirations and judgements are given to us in the Jewish-Christian
religious tradition",[21] was not an adherent of any creed, and did not
believe in a personal God, in spite of often speaking of Him in personal terms, and wanting to know His thoughts and how He created
the world. At first glance, the Special Theory seems incompatible

---

[21] *Out of My Later Years*, Thames and Hudson, 1950, p.23.

with there being any omniscient Being, who could know immediately what was happening anywhere in the universe; omniscience would import a canon of absolute simultaneity that was contrary to the whole thrust of the Special Theory. That claim, as we have seen, is over-stated.[22] The existence of an absolute canon of simultaneity is not inconsistent with the Special Theory, but does result in there being a preferred frame of reference, and this is something we may be led to accept anyhow, in order to accommodate the background radiation or the general form of the General Theory.

Nevertheless, the Special Theory does not sit easily with Newton's unitarian theism. In particular, as we saw in Chapter 4,[23] the Parity of Esteem enshrined in the Radio Rule articulates a principle of projection which we employ in recognising other people as people and ascribing thoughts and feelings to them. Whereas in Newtonian mechanics there is only one point of view, in the Special Theory there are many. Whereas in Newtonian mechanics it is natural to regard space as a single, simple entity, and to think of it as God's *sensorium*, the Special Theory is more reluctant to take a God's eye view. Whereas in Newtonian mechanics God can be, and we can imagine ourselves being, in two places at once, in the Special Theory we can only project ourselves into another observer's frame of reference, and ask how things are from his point of view. Instead of seeing the universe as the uniform object of divine omniscience, we understand it according to a principle of universal empathy.

§8.8 Newton and Einstein: Relative and Absolute

The popular understanding of the Special Theory has been greatly darkened by the use of the name 'relativity' and the term 'relative'. The word 'relative' has many different meanings: it is always advisable to ask "Relative to what?" when the word is used, and to consider what exactly it is being contrasted with.

---

[22] §3.9.

[23] §4.6.

The Special Theory can be called a theory of relativity in that certain concepts—simultaneity, distance, duration, velocity, momentum, energy, acceleration and force—are relative to a frame of reference. Popular expositions have suggested that the Special Theory has relativised these Newtonian absolutes, and that now all is relative. But this is a facile conclusion. Although these magnitudes do, indeed, become frame-dependent, there are corresponding ones that are independent of the frame of reference. Spacetime separation and the energy-momentum four-vector are the same in whatever inertial frame of reference they are measured. Although there are certain samenesses that are preserved under the Galilean transformation but are lost under the Lorentz transformation,

Table 8.8.1   **How Absolute?**

	Simultaneity	Rest Frame	Universal Velocity	Space and Time	Efficacious
Aristotelian Physics	absolute	absolute	infinite	separate	yes
Mediæval Mechanics	absolute	absolute	infinite	separate	no
Newtonian Mechanics	absolute	frame-dependent	infinite	separate	no
The Special Theory	frame-dependent	frame-dependent	finite	integrated	no
The General Theory	frame-dependent?	frame dependent?	finite	integrated	yes

there are other, deeper samenesses that remain. The difference between Newtonian mechanics and the Special Theory is not that in the latter everything is relative to the frame of reference, whereas in the former it is absolute, but that as we move from the Galilean transformation to the Lorentz transformation the features that are preserved under these transformations are correspondingly altered. Once again, we find that neither Newtonian mechanics nor the

Special Theory is either relative or absolute: both are both; only, they differ in exactly what they take to be the invariant features of the physical world that do not depend on our choice of frame of reference.

No simple answer can be given to the question whether Newtonian mechanics or the Special Theory of Einstein is more relative than the other. All that can be done is to list the points of similarity and difference in Table 8.8.1 on the previous page. Once again, the entries in the last row are problematic. This is because the real contrast between the Special Theory and the General Theory is different from that between the Special Theory and its predecessors. Although Einstein was concerned in putting forward the General Theory to widen the equivalence class of phenomena that are to be treated as being essentially the same, there is a more significant shift in the aims the two theories seek to achieve.

In the General Theory gravitation is explained in terms of the variable curvature of spacetime. The General Theory seeks to give a geometric explanation of physical phenomena. The role of geometry in Newtonian mechanics and the Special Theory is quite different: geometry sets the scene, but does not itself explain anything. In Newtonian mechanics, as in its predecessors, we presuppose an Euclidean space, as the arena within which physical phenomena occur, but the geometry plays no part in explaining impacts or forces or the propagation of causal influence. Equally in the Special Theory we presuppose a Minkowski geometry of spacetime, which determines the limits of the possible, but does not seek to determine what actually happens. Minkowski spacetime, like Euclidean space, is not only of constant curvature, but of zero curvature; that is to say, it is flat. Its flatness indicates its low profile in the explanatory endeavours of physics. Geometry in the Special Theory, as in Newtonian mechanics and its predecessors, is kept very separate from physics. In the General Theory, however, the two are fused into a single geometrodynamics, in which the causal development of a physical system is seen as a geodesic path through a contorted spacetime. The integration of geometry with physics achieves great explanatory power, and not only explains the equivalence of gravitational and inertial mass, but explains gravitation without positing a mysterious force acting at a distance: but it is secured at a price. In undertaking to explain so much, it has to be able to explain the particular course of events of particular systems, and hence it has to become itself particularised. And particularity precludes generality. The spacetime of the General Theory is no longer homogeneous and isotropic, but varies in its geometrical properties from point to point, and at different points will have different geometrical properties, and different preferred directions. Instead of having a general geometrical back-cloth against which particular physical systems evolve on their own according to physical laws applying to their particular initial conditions, we have a single unified spacetime, a self-subsistent entity, somewhat resembling Spinoza's *Deus sive Natura*, evolving as a whole. The question of Absolute Time and Absolute Space begin to look different. Though locally they are frame-dependent, they are globally, in many solutions of the field equations, fixed. Hence the question marks again in the last line of the table. But really we are asking the wrong questions of the General Theory, and if we want to focus on the differences between the Special Theory and it successor, we should ask not about the absoluteness of time and space, but their *rôle* in causal explanation, and then we seem to have come full circle.

Disciples of Hegel will be quick to perceive the working of some dialectic in the way that opposing tendencies in successive physical theories are reconciled in a synthesis that it turn gives rise to a new antithesis. The Hegelian account is overdramatized, but there are differing, and sometimes opposed, aims of physical theory which make each seem in due course to be unsuccessful in certain respects, and thus to be ripe for revision. The requirements of unity, symmetry and togetherness cannot be completely reconciled. Newton pictured the material universe as a single space and a single time in which a large number of qualitatively identical point-particles were located, interacting by impulse only when they were in contact with each other. Space and time were each separately unified and highly symmetrical, and the point-particles, in being qualitatively identical, were symmetrical too, though many rather than one. But causality was difficult to account for in Newton's scheme. Interaction by impulse could not, as Boscovitch later showed, be accommodated within the Newtonian constraints of continuity and finitude: if point-particles really interacted only on impact, they would have to change their state of motion discontinuously, and so exert infinite forces on each other; otherwise each would have to act at least at an infinitesimal distance from itself, and we should be committed to some force acting at a distance or some force-field. Once, however, we take fields seriously, we are led by the arguments of Chapter 1 to the integration of space with time, and once we take causality as a continuous partial ordering, we are started on the approach of Chapter 3, and led to a Minkowski spacetime with its own intrinsic metric structure. In these respects the Special Theory is a natural development from Newtonian mechanics, giving greater weight to the need for a physical theory to account for causal interaction. Causal interaction is an aspect of the togetherness of things, and hence of the unity of the universe, but to the extent that it explains one thing in terms of another, it detracts from each thing's independence from everything else. Leibniz was a more consistent pluralist than Newton, and Newton than the later Einstein. The greater stress we place on causal explanation, the more we make each thing relate to other things, and hence the more difficult for it to be exactly the same as another thing. Newton offers a theory of substance—fundamental things—in which each is qualitatively identical to every other, being numerically distinct only by virtue of occupying a different position in space at any one time from any other thing. This symmetry is lost in the General Theory where matter ceases to have an independent existence, and is merely a singularity in spacetime. There is only one fundamental substance in the General Theory—spacetime itself— and what we think of as things are construed as being merely aspects of spacetime, after the manner of Spinoza. Hence the particularity of the General Theory and the way in which it comes back to resemble proto-physics. But too much particularity runs counter to another facet of causality, which is always covertly universal.[24] In seeking causal explanations we are looking for samenesses and symmetries. In Newtonian mechanics and the Special Theory we achieve this to high degree in the background geometry, which is arguably as symmetrical, homogeneous, isotropic and flat as possible.

The ordinal approach of Chapter 3 offers a different perspective again. Newtonian mechanics has one "absolute" equivalence relation, simultaneity, whereas the Special Theory has none. The Special Theory has one absolute ordering relation, but so does Newtonian mechanics, the difference being that for Newtonian mechanics it is strict, whereas for the Special Theory it is a partial, ordering

---

[24] See J.R.Lucas, *Space, Time and Causality*, Oxford, 1985, ch 3.

relation. But, somewhat surprisingly, the partial ordering relation does more work than the strict one, so that although Newtonian mechanics has an absolute canon of simultaneity which the Special Theory lacks, the Special Theory has its metric given by its topology, and is to that extent the less amorphous and more absolute of the two.

§8.9 Historical Perspective

Newtonian mechanics was seen its time as a complete rejection of medieval science. Every age thinks of itself as having made a decisive break with the past, and overlooks the many thoughts and ideas it has taken over from its predecessors. In extending the Equivalence Principle to cover frames of reference in uniform motion with respect to one another, Galileo was taking a very important step forward, but he and the other thinkers of the Renaissance owed more to their scholastic predecessors than they were ready to acknowledge. To quote Duhem:

> The science of mechanics and physics, of which modern times are so rightfully proud, derives in an uninterrupted sequence of hardly visible improvements from doctrines professed in medieval schools. The pretended intellectual revolutions were all too often but slow and long prepared evolutions. The so-called renaissances were often but unjust and sterile reactions.

And again:

> The demolition of Aristotelian physics was not a sudden collapse; the construction of modern physics did not take place on a terrain where nothing was left standing. From one to the other the passage takes place by a long sequence of partial transformations of which each pretended to retouch or enlarge some piece of the edifice without changing anything of the ensemble. But when all these modifications of detail had been made, the human mind perceived, as it sized up with a single look the result of that long work, that nothing remained of the ancient palace and that a new palace rose in its place. Those who in the 16th century took stock of this substitution of one science for another were seized by a strange illusion. They imagined that this substitution was sudden and that it was their own work. They proclaimed that Peripatetic physics had just collapsed under their blows and that on the ruins of

that physics they had built, as if by magic, the clear abode of truth. About the sincere illusion or arrogantly willful error of these men, the men of subsequent centuries were either the unsuspecting victims or sheer accomplices. The physicists of the 16th century were celebrated as creatures to whom the world owed the renaissance of science. They were often but continuers and sometimes plagiarisers.[25]

The medieval philosophers, indeed, had opened the way that was ultimately to lead to Newton. They had refined the concepts needed for Newtonian mechanics and formulated the first and most basic principle of equivalence in their many discussions on the questions Aristotle had raised in the *Physics* about the nature of time, space, motion, and causal explanation. Galileo had studied Bradwardine, who first had formulated the principle that acceleration was the derivative of velocity with respect to time rather than with respect to distance.[26] The first Equivalence Principle, that of the causal inefficacy of space and time, though not fully thought through, was *the* great leap forward that made physics, as the search for the most general features of natural phenomena, possible. In comparison with that, the subsequent extensions, by Galileo to frames of reference moving with uniform velocity with respect to one another, and, by Einstein in the General Theory, to all accelerated frames of reference as well, may seem a mere follow-up of a breakthrough already achieved.

But that is to go to the other extreme. It is easy, with hindsight, to be unfair. Whittaker referred to Einstein's 1905 paper as one "which set forth the relativity theory of Poincaré and Lorentz with some amplifications, and which attracted much attention".[27]

---

[25] P.Duhem, *Les Origines de la statique*, p.91; translated by S.L.Jaki, *Uneasy Genius: The Life and Work of Pierre Duhem*, Martinus Nijhoff, 1983, p.387. The importance of medieval philosophy for the rise of science is shown by M.B.Foster, "The Christian doctrine of Creation and the Rise of Modern Natural Science," *Mind* 1934, pp.446-468; 1935, pp.439-466; 1936, pp.1-27.

[26] See Marshall Clagett *The Science of Mechanics in the Middle Ages*, University of Wisconsin Press, 1959.

[27] E.T.Whittaker, *A History of the Theories of Aether and Electricity*, London, 1951, 1953, vol.2, p. 40.

Although in each case there was great, and insufficiently appreciated, continuity with the work that had gone before, there was great innovation too: new concepts were put forward which organized old knowledge in a fresh and fertile fashion. We do wrong if we overlook the greatness of the men on whose shoulders Galileo and Newton and Einstein stood: but equally if, in our anxiety to emphasize the greatness of the predecessors, we deny that Galileo and Newton and Einstein were themselves giants.

Chapter 9
Relativity and Reality

§9.1 Shaking the Foundations

The theories of relativity have often been thought to have a bearing on philosophical method. Some presentations have seemed to support a positivist view, in which we attribute no reality to anything except our immediate observations. Einstein's critique of absolute simultaneity lent support to that view. "How can we tell," he asked, "whether two events are simultaneous or not?". Inspired by Mach, he denied any meaning to claims that two events were "really" simultaneous, unless there was some way of telling that they were. When later the Logical Positivists of the Vienna Circle propounded the "Verification Principle", that the meaning of a proposition is constituted by the method of verifying it, the rise of the Special Theory was taken as vindicating it, and as showing how much better science progressed when freed from metaphysical lumber and concentrating solely on observable measurements.

Einstein's general stance seemed to support this interpretation. He acknowledged a great debt to Mach, praising his *Die Mechanik* in 1909 and writing to him about his "inspired investigation of the foundation of mechanics" in 1913. Mach endorsed Einstein's Special Theory in the second edition of his *The History and Root of the Conservation of Energy* and said that the latest advances of physics were turning into reality his often expressed view that "the foundations of physics may be thermal and electric".[1] Later he was at pains to resist Planck's absolutist interpretation of the Special Theory and insisting on the positivist interpretation, which Einstein himself was generally believed to support.

Einstein's unconventional appearance and behaviour conveyed the impression of a rebel, of one who had swept away the certainties of Victorian physics. Relativity came to symbolize the break-up of the old order, and seemed to show that even in the most exact of

---

[1] Quoted by S.L.Jaki, *The Road of Science and the Ways to God*, Chicago University Press, 1978, p.182.

all the sciences the old certainties had crumbled away. To many it was but a short step to say that all is relative, that there are no fixed and firm truths. All our stable and hallowed landmarks had been washed away, leaving only a vague subjectivism.

§9.2   Einstein's Intellectual Progress

Einstein's actual intellectual commitments were different from those he was credited with. In truth Einstein was never committed whole-heartedly to positivism, and moved further and further away from it in the course of his life. Whereas for Mach the world was to be reduced to his own *ego* and sensations, for Einstein the world always loomed large in its own right.[2] As early as 1901 he had written: "As regards science I have got a few wonderful ideas in my head which need to be worked out in due course. I am now almost sure that my theory of the power of attraction of atoms can be extended to gases.... It is a magnificent feeling to recognise the unity of a complex of phenomena which appear to be things quite apart from the direct visible truth." Einstein gradually came to realise that, though like Mach he had been captivated by Kant's *Critique of Pure Reason* as a young man, unlike him he had not remained its captive, and that his real views about science in general, and in particular the way the Special Theory should be interpreted, were diametrically opposed to those of Mach. In 1922 he described Mach's method as one that would provide a catalogue but not a system, and said that Mach was a good mechanic but a deplorable philosopher, and that his approach could never give birth to anything living but only exterminate harmful vermin. In later years he repudiated positivism more strongly and endorsed Planck's view of physical laws as describing "a reality in space and time that is independent of ourselves",[3] and complained to Schlick that his presentation of physical theory was too positivistic:[4] the

---

[2] S.L.Jaki, *The Road of Science and the Ways to God*, Chicago University Press, 1978, p.184, from whom the next two quotations are taken. See also his *The Absolute beneath the Relative,*, University Press of America, 1988.

[3] P.Frank, *Einstein, His Life and Times*, London, 1948, p.215.

[4] Quoted by G.Holton, *Thematic Origins of Scientific Thought: Kepler to Einstein*, Harvard University Press, p.243.

aim of physical theory was to find out not only how nature's trans-
actions are carried out but also why nature was exactly the way it
was and not otherwise.In his autobiographical notes he recalls "In
my younger years Mach's epistemological position ... influenced
me very greatly, a position which today appears to me to be essentially untenable".[5] It was not possible to construct science by the
ordering of sensations: "The mind can proceed only so far upon
what it knows and can prove. There comes a point where the mind
takes a higher plane of knowledge, but it can never prove how it
got there. All great discoveries have involved such a leap."[6] Far
from witnessing in favour of positivism, Einstein himself showed a
commitment to there being some sort of reality which it is the scientist's aim to discover and to understand. Though influenced by
Mach's positivism, he came to repudiate it, and thought of himself
not as simply following after sense-experience and cataloguing it,
but as aspiring to a God's-eye view of the world: "I want to know
how God created this world. I am not interested in this or that
phenomenon, in the spectrum of this or that element. I want to
know His thoughts; the rest are details."

§9.3 Verificationism

Einstein's authority, though weighty, is not conclusive. We need
to look at the arguments for verificationism and assess the Special
Theory in the light of these, before deciding whether the Special
Theory supports a realist or an anti-realist view of the world.

There are many different arguments for verificationism. The
most telling arise from persistently pressing the question "How do
you know?". The scientist is properly proud of basing his assertions
on empirical evidence, and if asked the question, feels bound to answer it. But then, after he has cited the experimental evidence that
would ordinarily be adduced in favour of his assertion, the further
question is asked why that evidence should be evidence in favour
of the conclusion it is being cited to support. In the normal interchange of physical argument the further question can be properly

---

[5] P.A.Schlipp, ed., *Albert Einstein: Philosopher-Scientist*, Evanston: Library
of Living Philosophers, 1949.

[6] Quoted by R.W.Clark, *Einstein*, Hodder and Stoughton, 1979, p.582.

asked and answered, and additional empirical evidence support-
ing background assumptions can be given; but if the question is
thereupon repeated and raised again and again, the stage will be
reached when no more empirical evidence can be adduced, and the
inference from the evidence to the conclusion must be an *a priori*
one. If it is maintained, as it was by the Logical Positivists of the
Vienna Circle, that there are no valid *a priori* inferences other than
purely deductive ones, it then follows that the conclusion the scientist infers from his empirical premises can contain no more than
what those premises themselves contained, so that their meaning
is just the empirical evidence on which they are based. Hence, it is
argued, though a scientist may think that his concepts go beyond
the empirical grounds that would justify their use, this is an illusion, and in so far as he is thinking cogently he can only be validly
claiming what is warranted by the empirical evidence alone.

There must be something wrong with this line of argument, for
if it were valid, the honest scientist could never be wrong. If all
he was asserting were the empirical data on which his theory was
based, then provided he had reported them correctly, no further
evidence could contradict what he said. But scientific theories can
be refuted by further evidence. Therefore their content is more
than the earlier evidence on which they were based.

It is right to ask "How do you know?", but wrong to press the
question too insistently. If I make a claim to know something, I
invite the question, and ought to be ready to say what my warrant
is. But the Logical Positivists were in error in supposing that the
content of my claim was simply the warrant I had for making it.
Knowledge is impersonal, whereas any answer I give to the question
"How do you know?" is necessarily personal. How do I know that
there is a tree in the quad? Because I have seen it. But I am not
you or anybody else: my warrant cannot be your warrant or his
warrant. Yet our knowledge claims are the same in each case.

Crude verificationism must be rejected, then, because it rules out
the possibility of inductive argument and of interpersonal assessment. A scientist argues from particular observations to general
laws and future observations, and is concerned not just with what
he saw with his own eyes but with what any competent observer
could see were he to take the necessary steps. On both scores he is
going beyond the empirical evidence in the conclusions he draws.
He takes his experiment not only as a particular observation made

on a particular date in a particular place, but as being a typical instance of what happens, and as therefore being an indication of the way things happen and of what will happen on other occasions. He also takes his experiment not just as an experience that he happened to have, a set of sense-data he might tell his psychiatrist about or write up in a stream-of-consciousness novel, but as an apprehension by him of an interpersonal truth that could be apprehended by anyone else. The assumption of generality, and in particular of omnitemporality and interpersonality, are assumptions a scientist must make, or he ceases to do science at all. They are constitutive of the scientific enterprise. They are denied by the assumptions implicit in the arguments adduced in favour of the verification principle, and to accept the latter is not to accept a new philosophy of science but to abandon science altogether.

The crucial distinction is that between what is claimed by an assertion and the warrant for making it. It was obscured for Descartes and the English Empiricists by their use of the word 'idea' to mean both a concept and a proposition, and Hume's analysis of cause is confused in consequence.[7] The force of the distinction can be best appreciated by considering assertions about other minds and other dates. I may know that you are in pain. I may know it by observing the gash in your leg, or by seeing you flinch, or by noticing the tense set of your jaw, or in many other ways: but none of these ways is equivalent to what I know, namely that you are in pain. It would never be self-contradictory to cite the evidence at my disposal and then to deny that you were actually in pain because of some defect in your nervous system, or because you were an exceptionally good actor well versed in simulating pains you did not really experience, or something. Yet it would be quite wrong on account of these logical possibilities to doubt that you were really in pain. A grisly counter-example arises from modern medical practice. For some sorts of surgery the drug curare is administered, as well as an anaesthetic, in order to paralyse the muscles and prevent any reflexes: and sometimes anaesthetics do not work. A patient could be fully conscious throughout an operation, and not be able to indicate to anyone the intense pain he was feeling. And if he died of the shock, nobody would ever know. There would be no warrant for asserting that he was in pain: but it would be a meaningful

---

[7] See J.R. Lucas, *Space, Time and Causality*, ch.3, pp. 27-28.

assertion none the less, and horribly true. What these possibilities establish is that the statement that some one is in pain is not the same as the evidence for his being in pain, but goes beyond it and claims something more.

Similarly there is a difference between the grounds for making a statement in the future or the past tense, and what it actually claims. If I claim that there will be an eclipse in 1997, you can check the grounds for my claim as well as I can by going over the relevant astronomical data. These are available now. Equally the evidence for statements about the past is evidence available now. In each case the warrant for the claim is quite different from the claim itself; the warrant is available now in the present tense, but the claim is not in the present, but either in the future, which is not yet accessible, or in the past, which no longer is.

Typically, in making a statement I stick my neck out, and say more than my grounds for making it. Although I should not properly understand the language if I had no idea of the circumstances that warranted my asserting a proposition, it is generally possible for my assertion to be warranted and yet mistaken. The meaning of the proposition is therefore not exhaustively constituted by the grounds for its assertion. In general terms its role in discourse is determined not only by the moves leading up to its assertion but those that can properly follow from it, and not only what it is to argue for it, and assert it, but what it is to argue against it, controvert it, and deny it.

More sophisticated verificationists consider not the actual evidence for, but the possible evidence for or against, an assertion. Instead of asking "How do you know?", they ask "What evidence would tell either way?". It is a good question, and one which the scientist should take to heart. By constantly pondering the question, he may be able to devise crucial experiments which will decide between rival hypotheses. Even if he does not achieve that, he may be able to pare down a theory and eliminate unwarranted assumptions. The verification principle enabled Einstein to free himself from the very strong and pervasive assumption of absolute simultaneity. Without Mach's teaching it would have been as difficult for him, as it was for his contemporaries, to question the absolute rightness of Newton's account. It is a good discipline for scientists to practise verificationism on occasion, on Fridays say, and test their theories to destruction. But unremitting destructiveness is arid and sterile. It is useful to try to devise an experiment to put some generally accepted assumption—that parity is conserved, for example—to experimental test; that way lie Nobel prizes: but it is wrong to draw philosophical conclusions from the failure to think up an experimental test of a principle that accords with physical intuition, or to suppose that if it is not susceptible of test it cannot be true; that way lies paralysis and uncompleted theses.

In recent years verificationists have based their case on the thesis that meaning is exhaustively constituted by use, from which they argue that we do not know the meaning of a sentence unless we can spell out the conditions under which it would be correctly asserted or refuted. But we often ask questions we cannot answer. Although we should be open to criticism if our question were such that we had no idea of anything that could count for or against it, we cannot be fairly criticized for not being able to specify exactly the considerations which would count for it. In the ordinary way of speaking we should certainly not charge a scientist with having used words meaninglessly if he aired a speculation and only later saw reasons why it was false. All that is required for a question to be meaningful is that at some time there could be some consideration which bears on it, not that there should be a full formulation of all the considerations which would decide it conclusively.[8] Nor is the argument improved if it is couched in the terminology of the philosophy of mathematics. The word 'effective' is used there in a special sense. An effective procedure is an algorithm which will terminate in a finite number of steps. This is a far cry from what we normally mean by effective. We should not allow that a person effectively knew the meaning of a sentence if he started on an algorithm which, though terminating in a finite number of steps, would occupy him for the rest of his life, nor should we deny that he did if his use was not algorithmic. Moreover, if meaning were tied to effectiveness in the mathematical sense, since there is no effective procedure for deciding whether a particular procedure is itself effective, we should be led to the paradoxical conclusion

---

[8] For further discussion, see Richard Swinburne, "Verificationism and Theories of Space-Time", in Richard Swinburne, ed., *Space, Time and Causality*, Dordrecht, 1983, pp. 63-76.

that any claim that a sentence was meaningful was itself a meaningless assertion.[9]

There may be other arguments for verificationism depending on some theory of meaning. Philosophers are fond of putting forward theories of meaning. Usually they are very difficult to understand, the only intelligible part of the theory being that when we say something we think we understand, we do not mean what we meant to mean. The scientist is sceptical, and rightly so. If a theory of meaning makes out that we do not mean what we think we mean, although it is always possible that it is right and we are wrong, it is also possible that the theory of meaning has failed to take into account some evident feature of ordinary usage, and is in fact false. The onus of proof is on the person putting forward the theory of meaning. Unless he shows himself able to understand the meaning of words sufficiently well to express his meaning in clear and intelligible language, he does not deserve a serious hearing.

This robust attitude is likely to commend itself. But in the philosophy of physics there still will be doubts. Theories of space and time are peculiarly susceptible to verificationist critiques. This is no accident. For space and time have a high degree of symmetry. So far as Newtonian mechanics and the Special Theory are concerned, a mere difference in position or date or orientation makes no significant difference. But if differences of position, date, or orientation, make no difference, it would seem that they must be indetectible, and if they are indetectible, it must be meaningless to talk about them. We could not tell if the universe had started half an hour earlier or had been displaced four miles to the North, so it makes no sense to discuss whether this might have happened. And hence Leibniz and many others have argued that all we can intelligibly talk about are relations between dates, or points, or events; that is, durations, distances, angles, separations and relative velocities. These relational concepts the physicist can talk about: but absolute dates, positions, directions, or velocities, are chimeras, metaphysical nonsense that no self-respecting scientist can have any truck with.

It is clear that there must be something wrong with this argument. If it were valid, we could have no concept of symmetry.

---

[9] See further, Alexander George, "The Conveyability of Intuitionism", *The Journal of Philosophical Logic*, **17**, 1988, pp.133-156.

To say that something is symmetrical is to say that if we were to change it in a certain way—*e.g.* rotate an ice-crystal through 60^0— it would not make any difference; but in saying this, we are claiming that although it would not make a difference, there was some difference nonetheless. We are using different standards of difference. Difference, like sameness, is a triadic relation, and we need to say not only what is different from what, but in what respect.[10] In physics we want to say that in respect of the form of physical laws certain transformations, such as displacements or reorientations, make no difference, but in saying that much we allow that in some other respect the transformation does make a difference, indeed that the mere fact of being transformed itself constitutes a difference. In his correspondence with Leibniz, Clarke was envisaging a God's-eye view, in which God might have decided to create the Universe half an hour earlier or locate it in a different place.[11] There would be no difference as regards the form of physical laws, or anything else that man could know, but there *would* be a difference from God's point of view, which we can *understand* even though we cannot *know*. More generally, the concept of symmetry requires that there be more than one standard of difference, and so cannot require that both be totally discoverable. It is enough that we can say what the difference is, even though we cannot tell what it is in the particular case. Once we distinguish the different sorts of difference implicit in the concept of symmetry, we can ask what sort of indetectibility the verificationist appeals to in his critique of space and time. If the world had been created a giga-year earlier, it would not have made any difference so far as the laws of physics are concerned, because only those correlations that are repeatable can count as laws of physics.

A further argument against verificationism from spacetime physics can be extracted from Whittaker's dismissive remark about Einstein's 1905 paper.[12] Historically, Whittaker was wrong. But, philosophically, his position would be tenable if verificationism were true. Lorentz had produced the Lorentz group, Poincaré the Equiv-

---

[10] See above, §2.1.

[11] *The Leibniz-Clarke Correspondence*, ed. H.G.Alexander, Manchester, 1956, Leibniz III §§5,6, pp.26-27.

[12] See above, §8.9, p.248.

alence Principle for the laws of electromagnetism[13] and the means
of calculating and predicting phenomena. If scientific theories were
merely formal systems or merely devices for describing and predicting observations, Whittaker's judgement would be a fair one. What
Einstein offered was not so much a set of new calculations as a new
Gestalt: instead of thinking of objects suffering a Lorentz contraction, we should see length itself altered, and correspondingly time
dilated, as we transformed from one inertial frame of reference to
another. It was a new view of what really was the case, and not
merely a convenient convention for talking about relations between
the motions of ordinary material bodies,[14] and it was on this account that it was so revolutionary and so important. Whittaker
was wrong because science is not so limited in its concerns as verificationist philosophies of science make out, but is concerned to
give an account of reality which goes beyond a mere description of
phenomena. A scientist is concerned not just to tell us how things
appear, but how they really are, and why.

§9.4 The Marks of Reality

It is difficult to know what we mean by the word 'real'. Originally it was a philosophers' word, coined by the Schoolmen from
the Latin *res*, meaning 'thinglike'. It was rapidly appropriated to
common use, and has become, like the word 'good', a chameleon
word, taking its colour from its surroundings. There is no one
property of realness possessed by real butter, as opposed to mar-
garine, real silk, as opposed to nylon, real numbers, as opposed to
rational numbers or complex numbers, or real estate, as opposed
to stocks and shares, any more than there is one single property of
goodness possessed in common by good apples, good news, good
intentions and good views. In each case the words 'real' and 'good'
pick out some members of a class from others possessing the con-
trasting property, but the criteria of selection vary with what is
being selected, our normal purposes and our normal expectations.
These criteria are often unclear and may diverge, and there is no
guarantee that we shall be able to find a single order of reality or
excellence.

---

[13] H. Poincaré, *The Value of Science*, Dover, 1958, p. 98.

[14] H.Poincaré, *Science and Hypothesis*, London, 1905, ch.VI.

When we consider the ontological status of natural laws and physical entities, we have several different criteria of reality. In the first place, what is real is contrasted with what is subjective, what depends only on my egocentric imagining or wishing. There is some sense that what is real is common to all persons, interpersonal, or invariant under change of person concerned. This is borne out by a criticism Robb made of Einstein's account of simultaneity:

> In particular, I felt strongly repelled by the idea that events could be simultaneous to one person and not simultaneous to another; which was one of Einstein's chief contentions. This seemed to destroy all sense of the reality of the external world and to leave the physical universe no better than a dream, or rather, a nightmare. If two physicists A and B agree to discuss a physical experiment, their agreement implies that they admit, in some sense, a common world in which the experiment is supposed to take place.[15]

The core of Robb's argument is that what is objective must be common to different observers and hence the same for them all. It is a valid contention, and his criticism of Einstein is met by distinguishing observers from frames of reference,[16] and arguing that despite appearances the Special Theory does not deny that there is something common to all observers, but merely specifies it differently, and more profoundly. The Special Theory is not only the same for all persons, but covariant under change of date and position and orientation, and furthermore under change to a different frame of reference moving with uniform velocity. In this it fulfils the scientific ideal as a continuing attempt to understand the world and to identify those quantities that remain invariant and those forms that remain covariant in the midst of the changing flux of phenomena. Thanks to the Special Theory we see reality more clearly than before, because we have identified more closely the invariant realities behind our various observations. Instead of absolute date, defined by absolute simultaneity, and absolute duration, we have absolute spacetime separation and proper time. We may, as Robb complains, have lost the d*t* of Newtonian mechanics,

---

[15] A.A. Robb, *The Absolute Relations of Time and Space*, Cambridge, 1921, Preface, pp. v-vi.

[16] See above, §2.5, p.45; and ch.4, §4.3, p.137.

but have gained instead the deeper $d\tau$. The dilation of time is a perspectival effect much like the elliptical appearance of a penny seen from an oblique angle; and as we learn to talk about the round penny rather than the ellipses each of us actually perceives, so we should adopt the four-vector approach of spacetime physics instead of the three-vector approach of schoolboy mechanics. Seen in this light the Special Theory is not in the least anti-realist, but on the contrary a great stride towards discovering the underlying structure of reality.

Reality is contrasted not only with what subjectively appears to me, but with what I subjectively aspire to or want. Hence a second mark of reality is that it is ineluctable; it exists independently of me, and will continue to exist whether I will it or no. It is potentially recalcitrant to my will, and could force itself upon my attention in spite of my wish that it should not. Reality is something I cannot wish away.

Reality is independent not only of me and my will, but of us and our wills. I cannot wish it away, you cannot wish it away, he cannot wish it away, and so neither can we wish it away. A third mark of reality thus emerges, which goes beyond that of mere inter-personality, omnitemporality, and holding for all points of view, and beyond that of recalcitrance to just my will. It is not only independent of me, but independent of us. We together can often establish conventions, understandings, or laws, which I, you, and he, must acknowledge: but we cannot stipulate that reality should be such and such; it is for us to conform to reality, not reality to conform to us. From there follows a fourth mark of reality that all our views about it, however widely shared and interpersonally endorsed, being corrigible with respect to what really is the case. It is always possible to side with Galileo and against the consensus of informed opinion. A scientific proposition may be mistaken, and scientists have often been shown to have been wrong. Science, therefore, is not just a cultural creation of scientists. Although it has, indeed, resulted from the aspirations and achievements of scientists, their attitude has been one of trying to be guided by the nature of reality, not just creating it as a social artefact.

The sense of ineluctability is mediated in many ways. It may be simply a brute fact, that no matter how hard we try, we cannot bring about one state of affairs without another following thereupon, but often it is a rational necessity we can understand which confers the sense of ineluctable reality. Often, then, we pick out as real what is most fundamental in our apprehension of reality. It may be fundamental because it is irreducible— certainly if some concept or theory can be reduced to another, it seems derivative, subordinate and less real, and if, *per contra*, we cannot give an adequate account except in terms of some concept or by invoking some theory, we are thereby led to accord it high ontological status. A further mark of reality is that it should be complete in itself. Einstein took this to mean that it should be completely determinate, and used it as an argument for the incompleteness of quantum mechanics, that being indeterminist, it could not be giving us the last word on the nature of reality.

Our intimations of reality apply to concepts as well as natural laws and theories. We regard them not simply as tools or constructs of our own, and our colleagues', devising, but as successive attempts to grasp the real underlying nature of things. It is because of this that the concepts of the Special Theory and Newtonian mechanics are commensurable. If concepts are simply to be defined in terms of the operations used to measure them or the theories of which they form a part, as operationalists and Logical Positivists maintain, then the successive changes of experimental technique and of theories employed would indeed result in entirely altered concepts which would have nothing in common except their name. On this view Newtonian mechanics and the Special Theory are simply incommensurable, because the fundamental concepts of space and time are defined differently in the two systems.[17] Certainly the theories are different, operating with different basic concepts, and seeking a different type of explanation. Even where the theories use the same terms, the concepts are not the same. The components of the energy-momentum four-vector in the Special Theory are not the same as the scalar energy and the three momentum vectors in Newtonian mechanics. It is then argued that the concepts used to describe any experiment are "theory-laden", and therefore there can be no way in which two competing theories can be distinguished, because any experiment designed to choose between them must be described by terms with different meanings.

It is clear that there must be something wrong with this argu-

---

[17] B. Barnes, *Thomas Kuhn and Social Science*, New York, Macmillan, 1982, p.65.

ment. A knock-down refutation is due to Allan Franklin.[18] Consider the collision of two balls of equal mass, one of them initially at rest. If the collision is elastic, Newtonian mechanics predicts that the directions of motion of the two balls after the collision will make an angle of 90^0 with each other, whereas the Special Theory predicts an angle of less than 90^0. Since the angle can be measured in theory-neutral ways, we have a definite way of distinguishing between the two theories.

The essential point is that 'neutral' is a relative, not an absolute, term. Although there are no absolutely neutral measurements —to measure an angle presupposes a certain theory of space and rigid motions within it—it does not follow that there are no experiments or measurements that are neutral as between Newtonian mechanics and the Special Theory. Time and again physicists have been able to perform a crucial experiment which decides between two theories. The experiment is theory-laden, but the theory it presupposes is not either of those being put to the test.

Although Newtonian mechanics and the Special Theory are different, they are not totally different: admittedly, the components of the energy-momentum four-vector in the Special Theory are not the same as the scalar energy and the three momentum vectors in Newtonian mechanics. But they are developments of it. As in other branches of science, our concepts develop with the growth of our knowledge. Initially our concepts are defined with a range of attributes, As our experience develops, we find that some attributes are modified, that new ideas are introduced, and that new relations are found with associated concepts. We keep the same name, but the meaning progressively shifts as science develops. The transition from the one to the other is not a sharp, unbridgeable one. The very fact that we use the same name argues against that. Moreover, in spite of the differences between the theories, the similarities are much greater: not only are the theories very similar in general outline, but under some conditions they approximate to one another. As $v \rightarrow 0$, the Lorentz transformation tends to the Galilean transformation; the General Theory approximates to the Special Theory on a sufficiently small scale. We are entitled therefore to make comparisons between the corresponding concepts of the theories, and see one as an approximation to, or refinement of,

---

[18] Allan Franklin, *The Neglect of Experiment*, Cambridge, 1986, p.110.

the other. Although we should, indeed, be sensitive to the nuances of different theories, and not blindly assume that because the same words are used, they must bear exactly the same meanings, and recognise that their meaning is constituted in part by the role they play within the theory, the theories are not isolated entities, but are related both to each other and to other theories and to empirical phenomena. These constitute enough of a common standard to allow some degree of commensurability, not always exact or complete, but sufficient to enable rough and ready comparisons to be made.

The fact that concepts can evolve is an argument for their referring to something real. In continually refining and changing our definitions we think of them as referring to the same object. This makes sense if we are trying to understand an objective reality existing apart from ourselves, unaffected by whether we understand it or not, and yet in some degree open to that understanding. The essential difference separating the anti-realist from the realist, is between referring and defining. It is possible to refer to a reality that is mediated to us by a scientific concept without possessing even a partial definition of its essential nature.[19] Hence it is that the many changes of definition, as our conceptual understanding has developed and deepened, do result in our talking at cross purposes in entirely different modes of discourse, but are evidence of our gradually improved apprehension of an underlying reality.

§9.5 Reality

Different approaches to the Special Theory yield different intimations of reality. Einstein's original approach has the most economical ontology. It starts with physical phenomena, located by means of some suitable frame of reference, and the laws of nature that correlate them. The Equivalence Principle de-emphasizes the importance of any particular inertial frame of reference, and focusses attention instead on the laws of nature themselves. They are what remain the same whatever frame of reference we adopt. Underlying correlations are underlined, while appearances that hitherto had seemed important are discounted as mere perspectival effects. Spacetime itself is not very real on this view, though since spacetime separation and the speed of light are the same in all inertial

---

[19] J.M.Soskice, *Metaphor and Religious Language*, Oxford, 1985, ch.VII.

frames and of fundamental importance for the laws of physics, some aspects of spacetime are indubitably real.

The Communication Argument fits a less economical ontology, but requires a less demanding extension of it. Frames of reference are the fundamental entities, Leibnizian monads, but not windowless. Each is both an observer and an observed, and the Lorentz transformation are the result of harmonizing these two roles under the condition of all observers being on a par. The thrust of the argument is intersubjectivity and rationality. We start with a quasi-Leibnizian ontology of persons, who can both observe, and communicate with, one another, and then consider how their schemes for referring to events must be co-ordinated if parity of esteem is to be established. It is an interpersonal, not impersonal, reality that is sought, one in which I am emphatically not the only pebble on the beach but in which we, together, are the arbiters of reality. It is an ontology in which I am individually humble, but we are collectively proud. If it be accepted that persons are a fundamental category of being, that they have minds and can see things from other people's points of view, and that they have bodies and can be seen by other people, then the minimum requirement of reality is that they adopt the Special Theory as their means of bringing their differing experiences into line.

In the integrationist approach of Chapter 1, spacetime is undoubtedly real. It is a fundamental part of our ontology, irreducible to a system of relations or set of natural phenomena. It sets the scene for all natural phenomena and is taken for granted in all laws of nature. But it remains always in the background, and plays no active part in the course of events. It is not one of the *dramatis personae*, as it is in the General Theory, but is always causally inefficacious, though in another way causally opaque.[20] The absolute location of any particular event is a matter of happenstance, which cannot be explained further, but must simply be accepted. It is not, however, a very profound fact: it is, along with absolute orientation and velocity, a factor we discount in framing causal laws or explanations of events.[21] This confers on spacetime a certain air of unreality which impels many thinkers to deny its fundamental role in our conceptual structure. But that is a mistake. The

[20] See above, §§8.5-8.6.

[21] See above, §8.5.

tenuousness of spacetime is in part stipulative,[22] and anyhow incomplete. In part we secure the inefficacy of spacetime by ascribing any evident differences among effects to some causal factor other than spatiotemporal location; and in any case although absolute spacetime location is not of physical significance, relative spacetime location, that is to say spacetime separation between events, is. Nor is it only spacetime separation that is important, but the universal speed, c, and the metric and topological structure of the whole Minkowski manifold. Although the spacetime approach to the Special Theory casts spacetime in a humble *rôle*, as providing only the stage on which events happen without dictating the course they should take, its very pervasiveness makes it influential none the less. Its featurelessness, that is to say its high degree of symmetry, is a remarkable feature, and constitutes a subtle and cogent constraint on possible courses of events. Minkowski spacetime may eschew the overt geometrodynamics of the General Theory, yet yields the Lorentz transformation, and shapes the four-vectors that could underlie the conservation of electric charge and Coulomb's inverse square law: whatever disavowals are made on its behalf, it is undoubtedly real.

The great merit of the absolute approach of Chapter 3 is that it does not tangle with two-faced equivalence relations: it does not introduce structure which it then, in the name of some symmetry, discounts. The fundamental partial ordinary relation of causal influenceability is not easily accessible epistemologically, and the fundamental entities, possible events, are not economically available ontologically, but once granted, together with some reasonable assumptions of continuity and dimensionality, yield a thing that claims interpersonal validity and has an ineluctability vindicated by experimental observation, while possessing a high degree of rational transparency. In comparison with the integrationist approach, it is more topological, less metrical—though, surprisingly, it yields a metric and the concept of orthogonality. Whereas in Minkowski spacetime the light cone is a corollary of its metric and Lorentz signature, it is, on the absolute approach, the fundamental feature of spacetime, from which all the other topological and metrical features flow. It is reasonable to regard topological features as more fundamental than metrical ones; it is reasonable to

---

[22] See further below, §10.4.

attach some weight to the directedness of ordinary relations, and the consequent ability of the absolute approach to accommodate the anisotropy of time: it is reasonable, therefore, to regard the absolute approach as affording the most fundamental approach to the Special Theory, and on this score to reckon the relational structure of light cones as possessing the greatest degree of reality.

Chapter 10
The Rationality of Physics

§10.1 Rationalism and Empiricism

The ineluctability of the Special Theory is a rational, transparent one, not a brute opaque fact. But it is not clear what rationality is. Traditionally, rationalism has been contrasted with empiricism, and physics, like all natural science, is an empirical discipline. Yet—obviously—it is also a rational activity, in which the physicist is guided by rational considerations, as well as constrained by the hard factuality of experimental observation. Both extreme rationalism and radical empiricism fail to do justice to the way physicists actually think. Plato was so confident of the superiority of theory that he was prepared to discount any evidence that did not fit his theory as merely showing the inferiority of the world of sense-experience: but to follow that line is to cut science off from the world of nature altogether, so that it ceases to be natural science.[1] The logical positivists and their successors had a "thin" view of reason, and assigned a correspondingly greater role to empirical evidence. On their view a scientific theory is simply a set of axioms or postulates formulated in some logistic calculus together with some rules for assigning some empirical interpretation to some of the terms employed. The axioms or postulates entail various consequences whose empirical interpretation can be tested by experiment. If experiment confirms them, well and good; if it does not, something is wrong with the theory, and it must be revised or replaced.

It is evident that this account fails to fit the facts of scientific practice. Scientific theories, as they are operated by scientists, are not just sets of axioms in some formal logistic calculus, but are complex systems in which the rules of inference and rules of application are neither precisely formulated nor sharply distinguished. Empirical confirmation, though very important, is not all-important. The Special Theory was developed almost independently of experimental evidence, and was accepted for many years before decisive

---

[1] See J.R.Lucas, *Space, Time and Causality,* Oxford, 1985, ch.1, pp.2-3.

experimental evidence emerged in its favour. Furthermore experimental evidence against a theory is often not decisive. Rather than abandon an otherwise good theory we either attribute discordant results to experimental error, or invoke further hypotheses to explain away the experimental evidence, or make some minor modification to the theory to accommodate them. Scientific theories are neither internally so rigid nor externally so fragile as is made out. Internally, the inferences are not rigid deductive ones, but ones less sharply formulated, though still acknowledged as cogent by competent scientists: externally, scientific theories are partly, though not completely, insulated from experimental refutation by appeal to experimental error, unknown factors, or more profound formulation.

An adequate philosophy of physics must be both rationalist and empiricist, espousing a rationalism more chastened than Plato's in the face of recalcitrant facts, and an empiricism more enlightened and explanatory than that of the Logical Positivists. But whereas the extremes are simple to state and easy to embrace, the middle view is complex to articulate and difficult to defend.

§10.2　Embarrassment

As philosophers of science we should take the actual practice of scientists seriously. The derivations actually given by physicists are seldom strict deductions. They rest on assumptions, often not fully formulated, which can be challenged without inconsistency. There are many different assumptions involved in the different derivations of the Lorentz transformation, and in deriving Maxwell's equations from it. The structure of physical argument as actually adduced by physicists, is seldom that of strict deductive entailment

$$P, Q \vdash R$$

but some other form of inference, which we may symbolize by

$$P, Q \Vdash_{phys} R.$$

What is the force of this sort of argument? Deductive entailment can be, as we shall see more fully in the next section, defined in terms of inconsistency. Anyone who refuses to acknowledge the force of a deductive argument is thereby convicted of inconsistency, and is putting himself beyond the pale of intelligible discourse. What happens if some one refuses to acknowledge the force of a physical derivation? He does not contradict himself, but does let himself in for some sort of **embarrassment**. I can refuse to allow the simple derivation of the Lorentz transformation given in Chapter 2,[2] if I am prepared to make out that the relationship is not linear, but to do that would let me in for further awkward consequences. I can resist Purcell's derivation of Maxwell's equations, if I claim that potentials depend not only on position and velocity but on acceleration too.[3] But actually to maintain such a claim would be embarrassing. Whereas with deductive inference we are obliged to concede the conclusion, because once having allowed the premisses it would be inconsistent then also to deny the conclusion, here we are obliged to concede the conclusion, because once having allowed the premisses it would be embarrassing then also to deny the conclusion.

We can express this in terms of the logicians' concept of deductive entailment: we can say

$$P, Q \models_{phys} R \quad \text{iff} \quad P, Q, \neg R \vdash \text{embarrassing consequences.}$$

Thus if the transformation between one frame of reference and another moving at uniform velocity is not linear, then space and time are not homogeneous: if there are no scalar potential and no vector potential, nature would be much less integrated and uniform than we had expected: if they do not together form a four-vector, the integration of space with time is much less fundamental: if potentials depend on acceleration, the laws of nature are much more complicated than we had hoped. Of course, hopes may always turn out to be dupes. But to assume that they are dupes from the outset is to deprive ourselves of all hope of making progress in science. It is therefore reasonable to assume that our hopes are fulfilled, and to be rationally reluctant to abandon such assumptions except for very good reason. Rather than do that, we accept the conclusion of a physical inference as following from its premisses, even though it is not deductively entailed by them, and no actual inconsistency would result from denying the conclusion while still affirming the premisses.

---

[2] §2.7, pp.53-58; see also §5.2, 156-161.

[3] See above, §6.7, p.205.

§10.3 Deductivism

The argument of the previous section is an argument from practice, and so far as practice is concerned, incontestable. But physicists still are drawn to pure deductive argument as an ideal, and are inclined to feel that their actual practice falls lamentably short of what it ought to be, and should be taken as something to be repented of rather than a guide to be adopted for the future. They therefore attempted a rational reconstruction of scientific argument in terms of deductive inference, and were impelled to suppose that this was how science must be if it was to be rational, and that in so far as scientists did not operate according to the canons of their reconstructed science, they were failing to be proper scientists.

The view that deductive argument is the only really valid form of reasoning is an old one, and has much to commend it. Deductive reasoning is maximally explicit, maximally coercive, minimally holistic, minimally hand-waving, minimally personal. It is maximally explicit. Any other form of reasoning can be converted into deductive reasoning by sufficiently persistent questioning. As we saw in the previous section, instead of saying simply

$$P, Q \mathrel{\Vdash_{phys}} R,$$

the physicist can be asked to spell out what exactly are the embarrassing consequences of refusing to concede R after having admitted P and Q. This is sometimes a useful exercise. It may bring to light hidden assumptions which had been hitherto overlooked. Much of the discussion in Chapter 5 was concerned with revealing the exact assumptions invoked in different derivations of the Lorentz transformation, and for this purpose we may need to cast them into deductive form. It is plausible to argue, then, that since it is sometimes a necessary step, it would be always a sensible safeguard. But that is to ignore the costs of total explicitness. The point of arguing is to be understood, and a totally explicit argument is difficult to understand: we cannot see the wood for the trees. Even in pure mathematics, which *is* a deductive discipline, complete formalisation is not a practicable ideal. Mathematicians, like physicists, need blackboards, so that they can rub things out quickly before they are copied down: the communicative power of hand-waving arguments is much greater than that of proof sequences. Although derivations play an important part in physics, they are not all-important. Einstein's great contribution in the Special Theory was not new derivations but new organizing concepts, a new *Gestalt*. The qualitative approach of Faraday provides the essential backcloth against which the quantitative Maxwell's equations make sense. And so, although it is always possible, and sometimes necessary, to cast a physical argument into deductive form, it does not follow that it is always necessary, nor even that it is possible always, to do so.

Indeed, we can go further. Even in mathematics, the ideal of absolute explicitness is unattainable, not just as a matter of practicable communicability, but in principle. To achieve absolute explicitness we should need to be able to spell out every rule of inference in some finite way, so that we can have a finite list of the "rules of the game" which we can specify before the game starts; only so can we throw the rule-book at any would-be objector. But we should not then be able to accommodate all of logic and mathematics. The logic which satisfies the requirement of being completely specifiable in definite finite terms —"First-order Logic" as it is called—is one in which we cannot define or categorically characterize the natural numbers, or give a definition of being finite or being well-ordered. We can give reasonable definitions of all these if we go into Second-order Logic, but then we lose finite specifiability, and find that however many rules of inference and axioms we specify, there are always further patterns of argument we can see to be cogent and further truths—well-formed formulae that are true under all reasonable interpretations—beyond those we can derive from the axioms we had specified according to the rules we had specified. So even in principle, as well as, manifestly, in practice, mathematics is not fully formalisable, and no absolute ideal of mathematical rigour exists that is adequate for all mathematical argument.[4]

---

[4] These results are corollaries of Gödel's theorem, which demonstrates that in any system of First-order logic which is consistent and adequate for elementary arithmetic there are well-formed formulae that are true but cannot be proved in that system. There is still, after fifty years, much dispute about the full significance of these results. Gödel's original paper was "Über formal unentscheidbare Sätze der Principia Mathematica und verwandter Systeme", Part I, *Monatschefte für Mathematik und Physik*, XXXVIII, 1931, pp. 173-198. tr. Kurt Gödel, *Lectures at Institute of Advanced Study*, Princeton, 1934; and B. Meltzer, Edinburgh, 1962 (but see review in *Journal of Sym-*

Deductive reasoning is maximally coercive. A person who fails to acknowledge the validity of an inference in a formal deductive system is simply breaking the rules of the game: a person who fails to acknowledge the validity of a deductive inference couched informally in ordinary language is guilty of inconsistency, which, if he persists in it, shows that he does not really understand the language he is purporting to use. In either case he has ruled himself out of court for serious consideration. Deductive arguments are those that a person must concede if he is not to put himself outside the pale of intelligible discourse. They are thus maximally coercive, and we can see why a philosopher, faced with a tireless sceptic, seeks deductive arguments to silence him. But they offer only over-kill to the physicist, who argues, not with philosophical sceptics, but with fellow physicists sharing not only a common language but common aspirations and assumptions as well.

Physicists are not sceptics. Although they are not credulous, and want to subject their colleagues' arguments to careful scrutiny, they see colleagues as fellow workers in a common enterprise, who share aims methods and beliefs. They do not have much time for the philosophical sceptic, and ordinarily need less blunt arguments than those required for making him see sense. Indeed, such arguments would not serve the different purpose of convincing a colleague. The sceptic has to concede the conclusion of a deductive argument on pain of self-contradiction, but the conclusions of interesting physical arguments are seldom ones it would be self-contradictory to deny. If a colleague doubts the conclusion I am urging on him, it is highly unlikely that he is being inconsistent in supposing the opposite of what I maintain. Deductive arguments

bolic Logic, **30**, 1965, pp.357-359). The expositions by P.J.Fitzpatrick, "To Gödel via Babel", *Mind*, LXXV, 1966, pp.332-350, and by Ernest Nagel and James R.Newman, *Gödel's Proof*, New York, 1958, London, 1959, are the best for the non-technical reader. The application of Gödel's theorem to the problem of mechanism and free will by J.R.Lucas, *The Freedom of the Will*, Oxford, 1970. has produced much controversy. For its intuitive aspect, see J.Myhill, "Some Remarks on the Nature of Proof", *Journal of Philosophy*, vol.LVII, pp.461-471; and Dale Jacquette, "Metamathematical criteria for Minds and Machines," *Erkenntnis*, 27, 1987, pp.1-17, esp.§7. For its bearing on the axiomatic development of physics, see T.A.Brody, The Axiomatic Approach in Physics, *Revista Mexicana de Fisica*, **27**, 1981, p.583.

fence the outer limits of intelligible discourse. The discourse of serious physical debate is much narrower. There are many theses which can be maintained without putting oneself outside the range of intelligible discourse, but which will none the less show that one is not discussing serious physics. If I put forward the proposition that on January 1st, 2000, everything that had been blue will become green, and *vice versa*, you can understand what I am saying, and may need to write a philosophical essay to show that I am wrong: but you would not think that I was doing physics, or that some argument from physics would be relevant to the discussion. In serious physical argument the uniformity of nature is taken for granted, as well as much else, and we reckon that these assumptions, together with certain aims and methodological approaches, constitute the discipline of physics, and seek for arguments within their compass when we want to convince a colleague.

Deductive reasoning is minimally holistic, minimally hand-waving, minimally personal. Scientists are sceptical about holistic, hand-waving, personal arguments, which may be all right in literary criticism, but are not, they feel, to be accounted scientific arguments. A scientific argument, like a scientific observation, is one that can be checked by one's colleagues. It differs from a presentation of a world-view as a whole, which I may by my eloquent gestures commend it to my colleagues in a congress, but where if someone is not persuaded, there is no obvious way of clinching the argument. In such a case, the unconvinced may well reckon that those who were persuaded were carried along by my flow of words or force of personality rather than the real merits of the case. The great merit of a derivation is that it is articulated into a number of separate steps, each one of which can be examined by itself to see whether there is any fallacy in it or whether it really is valid. And when the derivation is deductive, there is no covert pressure to avoid embarrassment, no need to have formed any *Gestalt*, to have "got it", to have seen the point of it all.

These are cogent considerations, which properly lead scientists to prefer step-by-step knock-down arguments where they can be had, and to eschew hand-waving arguments where possible. But they are not conclusive reasons for banning them altogether. After all, scientists want to communicate, want to explain, want to understand. Hand-waving arguments can often get across the point of a proof, where, notoriously, a formal proof sequence leaves most

people baffled. Organizing concepts illuminate and explain, where
sequences of formulae baffle and confuse. This was why Whittaker's
dismissal of Einstein's contribution to the Special Theory was ill-
considered as well as unjust.[5] We need to be able to see the whole
as well as the parts; we need to be able to explain as well as to verify; we need each to be able to achieve a personal understanding as
well as to press on others coercive arguments they cannot gainsay.
The contrast between a first-personal and intuitive kind of thinking, such as we have in literary criticism, and an omni-personal and
coercive kind, such as we have in mathematics and science, is not
a black-and white one. Instead of two polar opposites, we have a
spectrum of intermediate types. All serious thinking needs to be
both intuitive and coercive. The different types of argument are
complementary rather than mutually exclusive. Much as literary
critics justify their over-all interpretations by detailed examination
of the texts, so physicists illuminate their derivations by setting
them in the context of a general strategy. We need both to communicate our first-personal vision and convey the reasons we find
telling, and also to anticipate objections and avoid errors, making our argument not only persuasive to ourselves but convincing
to others. But these two aims are not always totally compatible,
and we need on different occasions to compromise one for the sake
of the other. Sometimes we use hand-waving arguments in order
to get our message across and enable our hearers to understand
what is biting us, and acknowledge that our arguments are fairly
open-textured and open to objection, but hoping that all serious
objections will be met in due course. At other times we go for
rigour, trying to block every loophole, and making every questionable assumption explicit, recognising, as we do so, that we shall
make many people unable to see the wood for the trees. The arguments of physics are towards the omni-personal coercive end of the
spectrum—more like those of mathematics than those of ethology
or taxonomy are—but do not have to be, indeed cannot be, at the
extreme end.

In our thinking about the nature of physical argument we should
not be distracted by the will'o'the wisp of absolute rigour. No such
ideal exists. Each argument can be formalised, if we so wish, in
order to make apparent exactly what assumptions and principles of

---

[5] See above, §8.9, p.248, §9.3, p.258.

inference are involved, but there is no exhaustive characterization of valid argumentation by reference to which every argument may be judged. Physicists should not feel guilty at not having pushed formalisation through completely: mathematicians do not do so either, and in principle cannot.

§10.4 Synthetic *A Priori*

The argument of the previous section has shown that physical arguments cannot be completely formalised, and in any case are not deductive because their conclusions can characteristically be denied without inconsistency. In Kant's somewhat awkward terminology they are synthetic *a priori* arguments, and the corresponding implications are synthetic *a priori* propositions. Kant asked "How are synthetic *a priori* propositions possible?" and himself answered the question with regard to causality by turning Hume's account on its head. Hume can be taken to have said that our idea of causal necessity is only in the mind, where it is the result of our being habituated to expect the effect on seeing the cause, as a sort of conditioned reflex. There is no causal connexion in the world: it is just projected on the world by the mind. Kant agreed that it was projected on the world by the mind, but instead of viewing this as just the result of our psychological make-up, maintained that it was the only way the mind could make sense of the phenomena it was experiencing. He went on to say that our ideas of time and space were likewise imposed by the mind in order to provide a coherent framework for experience.

It is illuminating to view the argumentation of the Special Theory in a Kantian light. The Equivalence Principle, the Principle of Reciprocity, the homogeneity of space and time, the requirement of covariance under the Lorentz transformation, and in the General Theory the general principle of covariance, are laid down as *desiderata* which we want physics to conform to.[6] Not only would we be pleased if it did, but we are prepared to make it.

In Chapter 6 we amended Ampère's equation so that it would hold even when the current is not steady, and stipulated that the laws of electromagnetism should be cast in a Lorentz-covariant

---

[6] Note in particular the stipulative form of Pars' derivation of the Lorentz transformation in §5.2, p.160; see also §2.4 pp.39-40.

form, and when they seemed not to fit, we massaged them until they did.[7] It is tempting, therefore to say that this is all there is to it, and that physics is just an exercise in the optative mood, whose fundamental guidelines are laid down by arbitrary *fiat*. Many philosophers of science have adopted this approach, and claimed that the principles of physics, in as much as they are not purely analytic, are simply conventions we have found it convenient to adopt. But that, as we saw in Chapters 2 and 4,[8] makes them too arbitrary, and denies the evident rationality of some of them: and accords ill with a subsequent justification when the Special Theory is seen from a different perspective.[9]

Even when viewed as stipulations, synthetic *a priori* propositions are not just arbitrary *fiats*, but reasonable requirements for which a rational justification, partly in terms of the aims, partly in terms of the presuppositions, of physical inference, can be given. It is reasonable to adopt the principle of natural uniformity: else inductive arguments are impossible, and we foredoom ourselves to failure in knowing about the natural world.[10] It is reasonable to hold that differences of person are *per se* irrelevant: else no account of an objective reality can be given. Equally, it is reasonable to hold that differences of date and position are *per se* irrelevant: else no experiments can be repeated. It is reasonable to discount the influence of remote factors, unless mediated through some intervening chain of causes and effects: else it would not be feasible to isolate causal factors from circumambient conditions. [11]

These are all reasonable principles to adopt, though none is logically necessary, and none need hold absolutely in all cases on pain of science otherwise being impossible. Nature is in many respects not uniform, and inductive arguments fail. We do take account of the person conducting an experiment or propounding a theory. Things were different at the time of the Big Bang. We may have to abandon the principle of locality in quantum mechanics. But

---

[7] See above, §6.6.

[8] §2.8, pp.59-61; §4.3 pp.135-138.

[9] See above, §3.4, pp.102-105.

[10] See J.R.Lucas, *Space, Time and Causality*, Oxford, 1985, ch.2, pp. 24-26.

[11] See J.R.Lucas, *Space, Time and Causality*, Oxford, 1985, ch.4, pp.56-61.

these are second-order modifications. Although we can compromise them in fine detail, we still need to reckon that they hold for the most part. Some experimentalists are better than others, but even the best needs to have his results reproducible by other competent workers. We accord great authority to the *dicta* of famous scientists: but even they can make mistakes. Though not all dates are exactly alike in all respects, most are in most, so that most experiments can be repeated, and most hypotheses checked. The principles of natural reasoning, though not absolute and not foolproof, are none the less generally reliable, and we should be foolish to jettison them if we had any ambitions to know the nature of physical reality.

The pragmatic argument is not only prospective, but retrospective. Although it is always possible that a synthetic *a priori* proposition, being synthetic, may be falsified by experiment, and although this has on occasion happened,[12] it remains substantially true that the physics based on these principles of natural reasoning has been vindicated by experimental results. It is because the massaging has been found to work that we think it tolerable to inflict it on an erstwhile respectable law of nature. It is like a lens, which may distort, but can clarify. The four-vector approach requires us to see laws differently and in a different light: it could be just eyewash, but in view of the greater grasp it gives, borne out in due course by experiment, we are reasonably confident that the spectacles we are putting on when we conform our thinking to some canon of natural reasoning are helping us to see more clearly reality as it is, rather than merely projecting on it creatures of our own imagining.

Although it is helpful to see the principles of natural reasoning as stipulations arising from the aims and presuppositions of science, and then justify them, in a Kantian manner, as the only possible guides for those who seek a certain sort of truth, we do not have to see them only as rules we lay down. Kant was impelled to do so, because he had circumscribed the role of reason too narrowly, leaving little room for propositions that were true *a priori* without being analytic. But no good reasons have been adduce for adopting so narrow a view of reason: indeed, none could be, if reason were so restricted as to allow only analytic propositions to be *a priori*

---

[12] See below, §§10.6, 10.7.

true.[13] We do not need to feel queasy about synthetic *a priori* propositions any more than about non-deductive inference, or feel that we can accommodate them only if they are to be regarded as being in the imperative, rather than the indicative, mood. In fact they have both aspects, and can be seen sometimes as the principles of natural reasoning, *rules* for reasoning aright about natural phenomena, sometimes as truths about the underlying nature of physical reality, the principles of natural philosophy, as we may call them.

§10.5 The Principles of Natural Philosophy

There are many principles of natural philosophy besides those mentioned in the previous section. The Schoolmen argued for the unity of nature on monotheistic grounds.[14] Besides unity, there are various ideas of sameness and symmetry, some of them implicit in the uniformity of nature itself, which are constantly invoked by the physicist, together with the pervasive assumption that the world is rational, and phenomena explicable. Modern physicists yearn for the integration of all their theories into one great unified theory. In Chapter 1 the underlying theme was the integration of space with time, and this principle was invoked in several derivations of the Lorentz transformation. Integration leads on to other concepts, not only unity and uniformity, itself clearly connected with some sorts of sameness, but to wholeness, coherence and togetherness, which is articulated by mathematicians as the concept of continuity, a concept often taken for granted, and sometimes explicitly invoked in the course of this book.[15]

Although the principles of natural philosophy are in some cases developed with formidable sophistication, the concepts involved are simple. Unity is very simple indeed. Sameness, as we saw in Chapter 2, is not so simple, being a triadic rather than a dyadic, relation, so that we need to specify the respect in which two things are the same; none the less, equivalence relations, being those that

---

[13] See J.R.Lucas, "Reason Restored," in Stephen Holtzer and William Abraham, eds., *Essays in Honour of Basil Mitchell*, Oxford, 1987, pp.71-84.

[14] Aquinas, *Summa Theologiae*, Ia q.47 art.3.

[15] For example in §1.1, §§3.2-3.4, §5.2

are transitive and symmetric, are of a very basic sort, as are also ordering relations—those that are transitive and asymmetric—which underlie the approach of Chapter 3. Whereas in Chapter 2 we had to consider sameness in respect of many different frames of reference, in Chapter 3 only one ordering relation, that of causal influenceability, was needed. Continuity seems more complicated, but can be analysed in terms of least upper bounds, themselves expressible by means of ordering relations.

We can sketch a tentative argument for the principles of natural philosophy not merely as methodological rules we adopt on pragmatic grounds, but as synthetic *a priori* truths. We start with the uniformity of nature, which not only legitimates inductive argument but expresses our conviction that the world is an orderly one where there are regularities to be discovered. Amid the welter of confusing phenomena there are some patterns to be discerned which enable us to predict what will happen and explain what has happened. Quite apart from the pragmatic unwisdom of forswearing inductive inferences, it is incoherent to do so if I have any notion of a reality other than myself I need to come to terms with, and unreasonable to do so if I have any hope of reality's being rational and capable of being understood.[16] If nature is uniform, there are some significant *samenesses*, and hence also *differences*. Differences, however important some may be, can never be absolute, or no explanation would ever comprehend them. If an explanation is to cover different phenomena, there must be some principle of unity between them. But that unity cannot be a simple uniform sameness in all respects, or there would be nothing to explain. The differences must be real, but not so absolute as to preclude their being integrated into a composite whole. If there is to be anything to explain, we must be pluralists, and to that extent agree with Leibniz rather than with Spinoza, but if there is to be any explanation, the many different things cannot be completely disparate, or harmonized merely by some pre-existent artifice, but must naturally cohere, bound together not by a Great Chain, but some more tenuous web, of being. Intimations of togetherness are more pre-

---

16 Much more can be, and has been, said about the justification of induction. See, briefly, J.R.Lucas, *Space, Time and Causality*, Oxford, 1985, ch.2; and more fully, R.G.Swinburne, ed., *The Justification of Induction*, Oxford, 1974.

cisely articulated in requirements of continuity, both as criteria of identity and as conditions of explicability. They also stem from our concept of time, itself rooted in wider considerations than those of natural philosophy alone. Once we have the concept of time, we can argue that it, and therefore also space and spacetime, must be continuous.[17] But in establishing time and space and spacetime as pervasive manifolds within which all natural phenomena can be integrated, we are in danger of differentiating them too much unless we match the requirement that all phenomena be spatiotemporally located with strong canons of irrelevance. Differences of time and space, we say, must be *per se* irrelevant, not only because otherwise we could not repeat experiments and check up on observations, but because time and space and spacetime are inherently too individuating: unless their individuating effect is blunted by some canons of irrelevance, everything becomes utterly different, and we lose the samenesses that are the object of our search. Hence homogeneity and the causal inefficacy of time and space and spacetime. But even as we say this, we say too much. We need some spatiotemporal criterion of relevance as well as of irrelevance if we are to distinguish between differently located but otherwise similar potential causal factors. It cannot be absolute spatiotemporal location, but there is no reason why it cannot be relative spatiotemporal location, that is to say spatiotemporal distance. Distance and duration are relevant causal factors, not ruled out, as too individuating, by any doctrine of the homogeneity or causal inefficacy of time or space or spacetime. We are led to a certain sort of relationism—causal relationism we might call it—not by some materialist doctrine that only material objects exist or doubts about the real existence of time or space or spacetime, but as the only available way of allowing spatiotemporal factors to enter into causal explanations in a form sufficiently limited to exclude complete arbitrariness.

If only differences between spatiotemporal locations, and not the absolute locations themselves, are causally relevant, it is natural to think that causal laws should be expressed in terms of differential equations. But this is a further step. No inconsistency is involved in the concept of action at a distance, whether spatial or temporal, only a difficulty in understanding it. Thus if we tried to make sense

---

[17] See further, J.R.Lucas, *A Treatise on Time and Space,* London, 1973, §§6,7.

   pp.29-42.

of telepathic communication,[18] it would be by seeing if its strength varied with the distance, or was affected by obstacles. If it was, we should begin to construct a theory of telepathic communication that took account of these constraints, and, for reasons that will appear shortly, would seek to cast it in wave form. If we failed, or if we found that telepathy was entirely unaffected by any physical circumstances, or if we failed to find any explanation of it, we might still have to accept it as a phenomenon, but should, like Leibniz, regard it as an occult power. Newton's law of gravitation was a case in point. The empirical evidence in its favour was decisive, even though it remained entirely mysterious how it could operate, so much so that Locke felt impelled to modify the fourth edition of his *Essay* and acknowledge that Newton's law of gravitation was well attested scientifically, though still maintaining our inability to conceive how it could work.[19] But though we may be forced to accept some law-like regularity involving action at a spatial or a temporal distance, we do not feel happy with it until we can trace some causal connexion across the distance, because we do not see how a cause could operate, except in its own spatiotemporal vicinity.[20]

The principle of locality is, however, more than an ideal. Not only can it be argued for pragmatically, as in the previous section,[21] on the grounds that we need some canon of locality to enable us to discount remote causes and concentrate on the local factors; but it is also a theoretical safeguard against the fallacy of internal relations. Everything in the universe is, in a manner of speaking, related to everything else. If I stretch out my little finger to the South, α-Centauri becomes that much closer to my little finger. But α-Centauri has not changed. Real, causal change is narrower than all describable change. Some factors cause real change, others do not. We need some further criterion to distinguish those relational factors that really are potential causes from those that are not. The only available spatiotemporal one is locality. It cannot be

---

[18] See above, §5.3, p.165.

[19] *Essay Concerning the Human Understanding*, bk. II, ch.8, §§11,12.

[20] See above, §5.7, pp.179-180.

[21] p.266.

absolute spatiotemporal location, already ruled out as being too individuating: it cannot be relative spatiotemporal location, because that excludes nothing. It is reasonable, then, to rule out action at a distance, in the absence of some mediate causal connexion, because by so doing we can distinguish the substantial causal relevance of local factors from the bare logical relatedness of all spatiotemporally located factors, however remote. Although it is not inconsistent to be an astrologer and to think that the stars in their courses affect human affairs without any intervening causal process, astrology offers no uniform way of picking out those conjunctions that do, from those that do not, influence sublunary events. It is more rational to seek a non-fortuitous criterion of causal relevance, and it is conceptually most economical to hold, as we did in Chapter 1, that causal influence is propagated along spatiotemporally continuous paths, and that the fundamental laws of nature are best expressed in the form of differential equations.

Locality also provides a criterion of identity. In Newtonian mechanics we think of material objects as possessing mass, and as occupying at any one time some part of space, and therefore as having shape and size, but not as having, in the last analysis, any other, secondary qualities. They have only primary qualities, and these are few and austere. They cannot characterize an individual very fully, certainly not completely.

There is therefore the possibility of many individuals, all qualitatively identical, though numerically distinct. Thus the physicist thinks of the world as composed of atoms, with all the atoms of a particular isotope of an element being exactly the same in all possible respects. The only difference between, say, two atoms of deuterium in the same quantum state is in their being differently located at any one time. Two deuterium atoms cannot be in exactly the same place at any one time. Difference of spatial positions at the same time constitutes the criterion of numerical distinctness. But although two material bodies cannot be in the same place at the same time, neither of them has to be in any one particular place. A thing can move. If it does, it will be in different positions at different times. How then can we tell whether any particular atom of deuterium is the same as one at a different time or different? The answer is by seeing whether it occupied a spatiotemporally continuous path between them in the interim. If it did, it is the same; if not, they are different. Spatiotemporal continuity is the only criterion left, granted the austere stock of primary qualities allowed by the Newtonian scheme. It would be different if we were operating with Leibnizian monads, each characterized by an infinite set of monadic predicates. Then no two individuals need be qualitatively identical. Any two distinct monads would differ in respect of at least one quality, which could be predicated of the one but not the other. Persons are like Leibnizian monads, and do not require, as a matter of logical necessity, spatiotemporal continuity as a criterion of identity. I can imagine myself falling asleep as a don in Oxford, and waking up to find myself an undergraduate in Cambridge, or a general in the Bolivian army, or a ruler of a one-party state in Libya. But atoms are less complicated than persons, form no intentions and have no memories, and so, having no resources for establishing their identity, are obliged to rely on locality alone.

The principle of locality has further implications provided we also adopt some conservation law for the propagation of causal influence. It then leads to there being more than one dimension of space, since only under that condition will the effects of a causal factor die away as they spread out from their original source. In a one-dimensional universe the argument with the astrologer would be only an empirical one. The stars could, and indeed would constantly, influence the course of terrestrial affairs. In two dimensions their influence would, in general, diminish with the distance, in three with the square of the distance, and so there is an *a priori* presumption, rebuttable in each particular case, but in the absence of rebuttal decisive, that the influence of the stars is negligible. The line of argument that leads us to posit the irrelevance of absolute and relative but remote spatiotemporal locations, while allowing the relevance of local factors, leads us also to look for the universe's having more than one dimension of space.

Physicists tend to take it for granted that the appropriate differential equation for the propagation of causal influence across a field is the generalised wave equation;

$$\frac{\partial^2 \phi}{\partial x_1^2} + \frac{\partial^2 \phi}{\partial x_2^2} + \frac{\partial^2 \phi}{\partial x_3^2} + \cdots + \frac{\partial^2 \phi}{\partial x_n^2} = 0.$$

This is arguably so, but, as we saw in §5.3,[22] it is not logically necessary, and should not be simply taken for granted. We need

[22] p.165.

to ask what is so special about the wave equation that if once we abandon impulse as the paradigm of causal interaction, waves seem the only alternative. In part, it is an issue of greater generality. Fourier analysis can accommodate impulses; even a bullet, provided it moves with finite velocity, can be analysed as a wave train: a wave account is not an alternative, and arguably a second-best alternative, to an impulse account, but a more general one, much as the Lorentz transformation is a more general version of the Galilean transformation. Moreover the wave equation clearly satisfies the Equivalence Principle in that it is a second-order differential equation concerned only with accelerations (that is to say the rate of change of the rate of change of the field) as the determinants of change. The wave equation satisfies also a strong principle of locality: the acceleration at a point depends only on the value of the field at that point and its average value in the locality. It has negative feed-back in as much as the acceleration at a point is towards the average value in the locality, and increases the more the value of the field at the point in question departs from the average value in the locality; so the wave equation is in a sense stable. But it is not too stable: it does not die away, because the value of the field at any point is always overshooting the average value to which it is' tending, and then needing to be corrected in the opposite sense. Thus the wave equation satisfies a number of *desiderata*, even though it has not been shown that it alone does. If it can be established that the propagation of causal influence through a field must be by waves, then we can make use of Narliker's derivation to justify the Lorentz transformation without appeal to linearity.[23] Thus the very strong locality condition implicit in the wave equation yields the same result as the condition that spatiotemporal location should be *per se* irrelevant.

There are further arguments from the propagation of causal influence. We need it to be sharp. In a one- or two-dimensional space the effects of a particular cause become blurred. For example, if a stone is dropped in a pond, it produces not one, but many concentric rings of ripples. In a three-dimensional space, by contrast, the wave-front remains sharp, so that a sharply localised initial state is observed later as an effect that is equally sharply delimited. This condition, known as Huyghens' Principle, is clearly

---

[23] §5.3, p.165, §5.4, pp.170-171.

needed if information is to be be transmitted from distant events. The communication argument of Chapter 4 obviously presupposes the transmission of information, and in any case the radar rule requires it. Huyghens' Principle is not satisfied by the wave equation in one- or any even-dimensional space, and Hadarmard has conjectured that it is satisfied only by the wave equation and then only in three- or higher odd-dimensional spaces. Courant has a further argument that the wave front will not in general be sharp in spaces of five or more dimensions.[24] If that argument is valid, the communication argument is valid only in three-dimensional space, and only in that space can remote causes be distinguished by their different local effects. More generally, the special status of the speed of light presupposes that wave fronts remain sharp and identifiable, not dissolving into a blur of wave packets moving with different speeds. If we regard the light cone as being fundamentally a sheath rather than a sheaf,[25] then indeed we should be committed to Huyghens' Principle, and hence to the three-dimensionality of space. Our concept of causality thus may impose surprising constraints on the dimensionality of space and spacetime: and if the only possible spacetime has $3 + 1$ dimensions, then the inverse square law must hold, for energy to be conserved, and the assumptions needed for the derivation of Maxwell's equations in §6.6 are shown to be less contingent and more built into the nature of things.

The principles of natural philosophy stem from an intimation of there being a reality, independent of us but to some extent knowable and comprehensible. It is a single reality—there is just one universe—but a complex one, comprising many different things that can be distinguished both in thought and in practice from other things, without being completely disparate. They are not, as Leibniz would have us believe, windowless monads, each entirely independent of all the others, but are, rather, integrated into a universe in which, in spite of individual differences, there are many significant similarities. The spacetime manifold furnishes the means both of individuating different things and of integrating them into

[24] R.Courant, in E.F.Beckenbach,ed., *Modern Mathematics for the Engineer*, New York, 1956, p.101; see also J.R.Lucas, *A Treatise on Time and Space*, London, 1973, §48, pp.245-246.

[25] See above, §3.1, p.86; Figures 3.1.2 and 3.1.3; see also §3.6, p.111, and §1.1, p.4, and §1.2, p.7.

a causally connected whole. And as we pursue the themes of integration, sameness, causal connectedness and harmonious communication, we are led to endorse the fundamental rationality of the natural laws physicists have, in fact, distilled from observation and experiment.

§10.6 Experiment

The deeply rationalist account of the previous section gives a one-sided view of physics. And though, in view of the irrationalism of much philosophy of science,[26] it is important to stress this side, it is, none the less only one side and needs balancing by a recognition of the crucial role played by experiments in physics. Experiment is the lifeblood of science, and though it goes hand in hand with theory, and is often guided and stimulated by it, it is easy for philosophers to think of experiments solely as confirming or refuting theories, and not recognise their part in forming the way physicists think.

Experiments are of many types, depending on the degree of sophistication of the science. In the beginning of science there is no theory beyond a general belief that the phenomena of the world are interesting and worthy of study. Many people noticed the peculiar properties of amber and lodestone, and were frightened by thunder and lightning, but it was a long time before these diverse phenomena were understood as manifestations of the electromagnetic field. This came about gradually, as people with a special cast of mind, who wondered what lay behind these phenomena, strove to make sense of them. They believed that there was a hidden order that could be discovered. Einstein has recalled in his autobiography how he was deeply impressed when as a small boy his father showed him a magnet. He realised that there is a deep and subtle order behind things, and was stimulated to try to find it himself.

We are all surrounded by so many different types of events that the first difficulty is the very profusion of material. We see balls bouncing, birds flying, flowers growing, and all these and many other day-to-day events become familiar in the sense that we generally know qualitatively what is likely to happen. In addition to these ordinary events there are sometimes things happening that

---

[26] See D.C.Stove, *Popper and After: Four Modern Irrationalists*, Pergamon, Oxford, 1982.

take us by surprise because they do not happen often. When we are young our first thunderstorm or snowfall is an object of wonder, but with repetition becomes familiar. We can still be surprised by rarer phenomena like waterspouts and earthquakes, but by this time we have probably read about them and in this sense they are not entirely unfamiliar.

How does science begin? If we followed Francis Bacon we would try to record everything we saw. We could fill notebook after notebook in a frenzy of activity, and even then we would succeed in capturing only a very small fraction of all our experiences. It would be a heap of information, but it would not be science, nor would it lead to any understanding. The prime necessity is thus to have some principle of selection. Out of all the phenomena we must choose a small subset as worthy of study. Thus we might decide to study how projectiles move through the air, or the life cycle of some bird or insect.

Having chosen what to study, the next stage is to find some regularity of behaviour. Science is not simply the description of what happens in a particular case: it must refer to what happens in a large number of cases. Underlying this is the recognition of an objective world that behaves in an orderly way, so that in the same circumstances things will behave in the same way. More deeply, that things have their intrinsic natures, and that their behaviour is a consequence of their intrinsic natures. If therefore we can understand these intrinsic natures we can then understand why they behave in the way they do in different circumstances. There is here the potential for a great economy of thought: we do not rest content with a minute description of all possible behaviour patterns but seek the key to understand them in a simple way.

At this stage there is a deadly trap for the would-be scientist. If he is of an imaginative nature, and if he has some knowledge of mathematics, he might think that he can intuit the nature of the thing he is studying. Mathematics can be developed by pure thought, so why not science also? This is the opposite error to that of Bacon, who thought that science can be developed by accumulating observations. Descartes tried to develop his science by pure thought, starting from the idea that the fundamental properties of all bodies are motion and extension. Nature, however, does not yield her secrets so easily, and Descartes, like Bacon, failed to build a viable science. This only comes by a complex interplay between observation and ideas, that gradually gives us some understanding of the world. It was Newton who pioneered the road between the empiricism of Bacon and the rationalism of Descartes, and thus laid the foundations of modern science.[27]

We start along the road to this goal by making careful observations and if possible measurements as well. This is rarely obvious. What measurements can we make on lodestone or lightning? The first requirement is to find by observation some feature of the phenomena that seems to be always present and behaves in much the same way each time. Is it repeatable, and so presumably an essential part of it, and not some quirk due to an extraneous cause? We can then see not only whether that feature is always present in definite circumstances, but whether it is still there when we start varying the circumstances in different ways. We might find, for example, that it occurs only within a certain range of temperatures, or when it is illuminated by light of certain colours. This is the exploratory stage, when we try to produce the phenomenon in as many different ways as possible.

The early experiments, by Hauksbee and Franklin and many others, were devoted to exploring electromagnetic phenomena in a qualitative way. What effects will a magnet have on this or that, or can we somehow capture the forces of a thunderstorm by flying a kite in the storm-laden sky? Gradually some ideas of magnitude appear. Is there some effect that we can use to estimate how strong the new force is? A gold-leaf is moved by electric charges, a needle is deflected by a current.

As soon as we find something that is moved by the phenomena, we have in principle a method of measuring it. How far does it move, and how fast? We can measure the position of the gold-leaf as it is charged, and the deflection of a needle. Sometimes it is more difficult: how can we measure lightning? Franklin observed sparks from the string of his kite, but real progress only came much later with the invention of the camera. As a result of his repeated observations the scientist comes to be familiar with the phenomena in a certain way. He comes to know what is the likely sequence of events, and what are the magnitudes of the effects observed at each stage.

---

[27] S.L.Jaki, *The Road of Science and the Ways to God*, Chicago, 1978, ch.6, "Instinctive Middle".

It soon becomes evident that observation on its own is not sufficient. We want to be able to create the phenomenon at will, to turn it on and off, to vary the conditions, to see what happens when we interfere with it in this way and that. The scientist construct apparatus to enable him to do this, and thereafter progress is much more rapid. He greatly broadens his experience of the phenomenon.

Musing on all this, the scientist begins to have some ideas about the underlying reality. Perhaps electricity is a subtle fluid that travels along a wire. This primitive model suggests further experiments. We are now moving from the stage of more or less random observations, the stage of playing with the phenomenon, to the stage of controlled experiments, guided by theory. At first the experiments are qualitative: will this happen if I do that? Will this attract or repel? Then it becomes possible to put some numbers to the strength of the observed effects. The battery is invented, and the electric current it provides makes systematic experiment much easier than before.

With the battery, Faraday makes a whole series of experiments and discovers the chemical effects of electricity. He develops a rather detailed understanding of electricity, still couched in qualitative terms. He thinks of electricity as a fluid, and the lines of force of the magnet as rather like rubber bands. He has the imaginative understanding of the experimental physicist that enables him to feel 'at home' with the phenomena. He knows, more or less instinctively, what will happen in various circumstances. It is like the knowledge of mechanics possessed by the tennis player, or of hydrodynamics by the swimmer.

Underlying it all is the familiar process of learning in physics, repeated by every student as he gains familiarity with the behaviour of some physical phenomenon. First of all he thinks about it using analogies with things more familiar. In this way he extends his knowledge of over some area of nature by integrating the particulars to grasp the reality and becomes more and more confident that he is well on the way to mastering the field. He is developing his tacit knowledge[28] of the phenomena, so that he can now answer with some confidence an indefinite number of questions about aspects of the phenomena of which he has no direct empirical knowledge.

---

[28] Michael Polanyi, *Personal Knowledge*, London, 1958, part 2.

As the measurements become more accurate, it becomes possible to look for, and to find, simple numerical relationships. The scientist expresses his model in mathematical terms, determines its parameters by comparison with measured data, and uses it to predict new phenomena. Ohm's law, Kirchoff's laws, Faraday's Law, Ampere's law, all embody relations between measurements in defined circumstances. At the same time the fundamental concepts of current, voltage, resistance, inductance are refined and ways of measuring them with ever-increasing precision are developed. Links are found between electricity and magnetism, and gradually a whole rang of apparently disparate phenomena are seen as manifestations of the same fundamental entities.

All this provides material for the theorist, whose aim is to unify all the phenomena into a tight mathematical formalism. Maxwell set out to do this, and started by reading Faraday's works. He tried to imagine the hidden mechanism behind the phenomena of electricity and magnetism. Following the prevailing beliefs, explanation was understood as reduction to Newton's laws. He devised complicated mechanisms with pulleys and strings and froth in his attempts to make sense of the laws of electricity and magnetism. Finally he arrived at a set of partial differential equations that bears his name and from these all can be deduced. It was then realised that the mechanical models were so much scaffolding that was no longer needed. The reality is the electromagnetic field, a new way of looking at the world.

Experiments are made for a variety of reasons: to obtain a more accurate value of some physical quantity, to decide whether one theory or another is correct, or simply to see what happens in an unexplored region of phenomena. Very often the experimentalist will have a fairly clear idea of what to expect; if he did not, he would not be able to arrange the experiment so as to detect it. Most of the time the experiment, if it works at all, gives a result within an expected range: occasionally a completely different result is obtained, and then the experimentalist's first thought is that it is just due to a malfunction of the apparatus, which indeed often it is; if, however, it is not, then he has to start again, to collect himself, to revise his ideas, his models, and his theories.

The development of science, and the contributions of theory and experiment to it, is reflected in the accounts of the work of the scientists. Early books and papers are full of descriptions of apparatus and of the phenomena they enable us to study. Later on there are numbers giving the results of measurements. Finally there is no mention of apparatus, and very few numbers, but only the austere beauty of partial differential equations. It is perhaps understandable that the philosopher of science, coming in at the end of the story, is most impressed by those equations. But a full account of science must give equal weight to the qualitative understanding of a Faraday, and the pioneer experiments of a Franklin.

This process of developing understanding continues in many areas of science; electricity, magnetism, light and heat are all understood by simple models and found to be governed by mathematical laws. With deepening understanding the scientist often realises that phenomena hitherto considered independent are aspects of the same underlying reality. Electricity and magnetism, for example, are found to be different manifestations of the electromagnetic field. Energy appears in various forms, electrical, mechanical, and thermal, and they can be converted one to another in precisely known ratios. We then begin to see that science is not just a series of disconnected subjects, but a unified whole. As our theories become steadily more sophisticated, Newtonian mechanics, Maxwellian electromagnetism, quantum mechanics, special and general relativity, grand unified theory and so on, we glimpse the goal of all science, the unification of all phenomena under a single all-encompassing theoretical scheme.

We are not quite there yet, but so much has been achieved that, in the physical sciences at least, we can now say that we have a rather detailed understanding of nearly all the phenomena we encounter, not only by simple observation, but also by sophisticated experiments that create conditions that occur nowhere else, in order to put our theories to the most searching tests imaginable.

§10.7 Surprises

The history of experimental physics is marked by a series of surprises. It has developed in ways that were totally unexpected at the time. Sometimes expectations were falsified. The Michelson-Morley experiment was not expected to yield a null result; Oersted thought that a current flowing perpendicular to the plane of the magnetic needle would not deflect it; Pauli was fairly confident that the result of Wu's experiment would be negative. So too, Rutherford had not expected alpha-particles to be scattered back from a thin foil—it was, in his graphic phrase, as if he had fired a 15 inch shell at a piece of tissue paper and it had come back at him. Equally unexpected was the inversion of the tip-top. Other surprises were not the falsification of predictions already made, but the observation of totally unfamiliar phenomena. Becquerel found his photographic plates fogged even though he had wrapped them in black paper and put them in a closed drawer; Hertz noticed a spark jump from one conductor to another at the other side of the room. The Arago spot in diffraction was surprising in a different way: it was a consequence of the wave theory of light, and was originally taken to be an argument against it, since it had never been observed. But then Arago looked for it and found it, thus vindicating the wave theory by surprisingly falsifying a generally assumed falsification.

Many of these surprises led to advances of great importance: the Michelson-Morley result, although of no importance to Einstein, was for many years the only experimental evidence that physicists could cite in favour of the Special Theory; Becquerel's work led to radio-activity; Hertz' to radio; Rutherford's to nuclear physics. The tip-top phenomenon, on the other hand, turned out to be deducible from classical mechanics. In that case the surprise was entirely due to our limited ability to follow out the implications of theories we had already accepted.

In other cases too the surprise may be due to our abilities being limited. It is recounted of Lord Cherwell that he "often said that, if only scientists had had their wits about them, they ought to have been able to reach Relativity Theory by pure logic soon after Isaac Newton, and not have to wait for the stimulus given to them by certain empirical observations that were inconsistent with the classical theory".[29] At first sight it is a startling claim, and seems to show the arrogance of the intellectual rather than the humility that is proper to the scientist. But if we think further, we see that the very fact that the scientist is sometimes surprised at the outcome of his experiments would itself be surprising unless scientists had usually reliable grounds for anticipating the outcome of their experiments. They are sometimes surprised because usually they are not, and correctly predict what will happen if a particular experiment is performed. The only way in which they could never be surprised would be for them to make no predictions, no generalisations and no extrapolations: in short, to abandon science altogether.

Explanation is linked with prediction. If I can explain why something is the case, I can see why it should be so, and thus have rational grounds for expecting it. If Lord Cherwell's retrospective verdict were completely inappropriate, it would mean that there was no rationale to the Special Theory: it was just simply a brute, opaque fact we simply had to accept but could in no way understand. But that is not so. And if, after the event, we can begin to justify the Special Theory on *a priori* grounds, these grounds would have been available for Lord Cherwell's scientist with his wits about him to predict the outcome of the Michelson-Morley experiment before it was actually carried out.

The extent to which surprises are due only to our inadequacies, and can always in retrospect be seen not to have been surprising at all, and discoverable by "pure logic", turns on the nature of reason, and what is meant by "pure logic". Once we recognise that there are many different types of reasoning, and that deductive logic does not comprise the whole of valid argument, we can use the fact that physicists have sometimes been surprised by the results of experiments to illuminate the account given in §§10.4 and 10.5 It shows, first, that the conclusions reached by physicists are synthetic: that is, they yield consequences which it is not inconsistent to deny— and which, in effect, were denied by Nature. It shows, secondly, that the arguments used by physicists are, in part, *a priori*, since, if the conclusion were nothing but a restatement of the premises, the only way a physicist could be surprised would be if he had made a mistake about his previous experimental data. Thus although it

---

[29] R.F.Harrod, *The Prof: A Personal Memoir*, London, 1959, p.57.

has often been held that synthetic *a priori* propositions are impossible, yet unless what was believed by the physicists was synthetic, they would not have been surprised at the outcome, and unless they had believed it *a priori* they would not have had any expectations in advance of the experimental observation to be upset by it. But how can valid reasoning be wrong? If a result, confidently predicted on the basis of physical theory, turns out otherwise than expected, something must go; and it seems difficult to allow that synthetic *a priori* propositions might turn out to be false, since if they are *a priori*, they are, according to Kant, necessary, and necessary propositions must be true.

The short answer is that 'necessary' is a "systematically ambiguous" term. As with the words 'possible' and 'impossible', there are many different senses of the word 'necessary', and what is morally necessary may not be legally necessary, and what is biologically necessary may not be physically necessary. So too within physics there are different modes of discourse, with correspondingly different modalities, and what is *a priori* and necessary with respect to one modality may be contingent and not necessary with respect to another.

The long answer requires a consideration of the response of physicists to experiments that disappoint their expectations. They do not immediately abandon the theory on which the prediction was based: theories are difficult to come by, and too precious to be lightly jettisoned. Sometimes an experimental finding is simply dismissed as an experimental error;[30] sometimes it is regarded as an anomaly, to be explained away in due course, but not casting serious doubt on the general truth of the theory. In other cases we do not discount the experimental evidence altogether, but rather than reject some fundamental physical principle, we call in question some other subordinate assumption, or modify the fundamental assumption so as to apply in a different way. Instead of abandoning the isotropy of space, we suppose that Oersted's experiment was not as symmetrical as it seemed, and that there was some factor which we had not taken into account.[31] Instead of just abandoning

---

[30] As happened when Miller repeated the Michelson-Morley experiment, and obtained a positive result.

[31] See above, §10.7.

the conservation of parity, we suppose that what is conserved is not just parity P, but TCP.[32]

If that were all physicists did, their *a priori* principles would be invulnerable, since any of them would be compatible with any counter-evidence—we could always save the fundamental principle by jettisoning some subordinate assumption—but useless, since we could never use them to make predictions in view of the total "jettisonability" of the subordinate assumptions. But the subordinate assumptions do not all stand on a par. We are much readier to abandon some than others. It is not a very deep principle of physics that all physical processes should be the essentially the same under time reversal T; indeed that principle does not apply at all in thermodynamics, and it is somewhat a matter of surprise that it should hold in Newtonian mechanics and electromagnetism. We should be readier to jettison T than P. Similarly, the principle of charge conjugation C is less deep than that of the conservation of parity P. If we had a choice, we should keep P and give up C or T, because parity expresses part of what we understand by the isotropy of space whereas charge is, so far as we know at present, less intimately connected with other fundamental concepts. In fact we did not have the choice, and there was no acceptable way in which the conservation of parity P could be saved in the face of Wu's experimental results; but rather than simply abandon the conservation of parity P, the conservation of something to do with parity, in the first instance PC and then, when that too was found not to hold, TCP was maintained instead. There is an order of jettisonability among physical principles. In the absence of counter-evidence, we expect them all to hold, and frame our predictions accordingly. If our predictions are not borne out by experimental observation, we scratch our heads, and wonder which of our assumptions was at fault, and suppose it must have been a very subordinate one. We may be able to devise an experiment to test whether it is, or may rethink our theory. If we can save our principle by jettisoning a very subordinate assumption, well and good. If not, we have to look for other, less low-level assumptions which may be wrong. What we *never* do is simply to conclude that the natural world is a strange place in which things happen for no rhyme or reason. We never jettison the principle of the uniformity of nature, or abandon the search for

[32] See above, §7.2.

natural laws and satisfactory explanations of physical phenomena. And we are almost, though not quite, as reluctant to abandon our cherished principles of the causal inefficacy of space and time, of their homogeneity, of causal continuity and the like; though each of these principles has had to be compromised in some branch of physics.

How exactly we order these principles it is difficult to say. What is essential is that there should be some order of priority, so that they can absorb counter-evidence without being effectively vacuous. The development sketched in §10.5 gives one such order, but it is not the only one, and sometimes we are forced by experimental evidence to revise our previous ratings of jettisonability. Although surprises can be absorbed, they cannot be ignored. And sometimes they open up a whole new field of physics. An experiment apparently goes wrong: the physicist falls down a small hole, and finds himself in a new continent.

§10.8 Rational Empiricism

The way that physics has actually developed reveals a constant dialogue between a rationalist and an empiricist strain of scientific thinking. Physicists need to live like empiricists, always submitting themselves to the verdict of experiment, but to reflect like rationalists, trying to make tidy sense of the deliverances of observation.

The interesting question arises whether the empirical element in the advance of science is inherent in the whole process, or whether it is just the unfortunate result of our limited powers of comprehension. If we had been more powerful reasoners, we would not have been surprised so often: if we had been infinitely powerful reasoners, would we have been surprised at all? In some cases the answer is clear. If we had had angelic intelligence, able to see in a flash all the consequences deductively entailed by any proposition, we would certainly not have been surprised by the tip-top, or by the Arago spot. We would have expected them to occur as deductive consequences of classical mechanics and of the wave theory of light. In other cases the answer turns on what precisely we mean by "reasoners". Deductive reasoning, we have argued, does not constitute the whole of reasoning, and physical inference is typically different from pure deductive inference. An angelic intelligence, capable of doing in a flash everything a computer could in principle do, could not have derived the Special Theory from previously accepted theories any more than it could have derived those previously accepted theories simply from the experimental observations on which they were based. For this some sort of inductive, rather than deductive, inference is required. But could an archangelic intelligence, capable of doing in a flash everything a computer could in principle do but capable also of further feats of inference of a rational but non-deductive kind, have been led to the Special Theory by ratiocination alone without benefit of empirical evidence?

Einstein could. Although according to the text-books he developed the Special Theory in order to explain the unexpected result of the Michelson-Morley experiment, he says in his autobiography that he first had the seminal idea when he was sixteen. He asked himself what a light wave would look like to someone travelling along with it. It must be represented by a stationary solution of Maxwell's equations. But no such solutions exist. So he went on to ask what transformation must be used to transform measurable quantities in one frame of reference to the corresponding quantities in another frame moving with respect to the first if the equations were to remain unchanged. It was the Lorentz transformation. And Einstein was led to it not by the pressure of empirical evidence but by that of rational necessity. Only if the Lorentz transformations were invoked would Maxwell's equations be the same from all those points of view which should be regarded as being on a par. "From the very beginning," he said, "it appeared to me intuitively clear that, judged from the standpoint of such an observer, everything would have to happen according to the same laws as for an observer who, relative to the earth, was at rest".[33]

Einstein's account differs from that of the text-books. Text-books over-simplify. The history of scientific discovery is complicated, and writers of text-books have no time to follow the actual course of discovery but only to present the results as clearly and simply as possible, and re-arrange the historical development to this end. They are concerned not with the workings of Einstein's mind, but with the reasons why we should accept the Special Theory as true, and the negative result of the Michelson-Morley ex-

---

[33] P.A.Schlipp, ed., *Albert Einstein: Philosopher-Scientist*, Evanston, 1949, p.53.

periment was for many years almost the only empirical evidence in favour (and as we have seen not always unambiguously available).[34] The text-books give not Einstein's own first-personal reasons why he came to think up the Special Theory, but omni-personal reasons why any scientist should adopt it. And of course, if these omni-personal reasons had not seemed cogent, and the Special Theory had not caught on, Einstein would not have related his own reasoning at length in his autobiography. Autobiographies are selective: we recount our successes, but pass over our failures in silence. First-personal reasoning is important, but not decisive: I may be guided by reasons far finer than those formulated by established orthodoxy, but whether they are good reasons is for others, and not me alone, to determine. But Einstein's testimony is decisive for our question. He was not surprised. He could have predicted the outcome of the Michelson-Morley experiment on the strength of *a priori* considerations alone, and was not shaken when Miller's contrary results were obtained at a later date. Neither was he surprised when the telegram arrived announcing the results of measurements of the bending of starlight by the gravitational field of the sun. He already knew, by reason of the General Theory, how it should behave. And yet it is still conceivable that the experimental result could have been otherwise. Indeed, around 1920 Einstein remarked that if only one of the three classic proofs of the General Theory were disproved, then the whole theory would be "mere dust and ashes"[35]

Are all advances of this type? Could Becquerel, Hertz and Rutherford have worked out by pure thought from what was known the results they observed? To say Yes is to make a large claim, unsupported as yet by any physical theory, and seems to deny that there is any real contingency in the world. But to say No is equally unattractive, in that it seems to impose limits to scientific explanation and deny the ultimate rationality of the world.

---

[34] See Michael Polanyi, *Personal Knowledge*, London, 1958, pp.12-13; J.L. Synge, *Relativity: The Special Theory*, Amsterdam, 1958, pp.161-162; or L.S.Swenson, *The Etherial Ether*, Texas, 1972. But see also A. Pais, *Subtle is the Lord*, Oxford, 1982, ch.6

[35] Statement made in a lecture recalled by H.Feigl, in R.H.Stuewer, ed., *Historical and Philosophical Perspectives of Science*, University of Minnesota Press, Minneapolis, 1970, p.9.

Once again the answer turns on different sorts of reason. If we take reason narrowly, as deductive reason, then the answer is clearly No. The surprising results of Becquerel, Hertz and Ruther-ford were not deductive consequences of facts and theories already known, which could have been worked out if only they were clever enough. In that sense there is certainly an objective contingency in the world. Things could be otherwise than they are, in the sense that it would not be inconsistent to describe them so. But there might be some other irrationality in so doing. Contingency is often contrasted not with deductive, but with rational, necessity. And then it is not obvious that there must be some objective contin-gency in nature (granted that there is a natural order at all), and the implications of supposing that there is are unwelcome. For any such contingency would be rationally inexplicable, a brute fact, which we just had to accept, but could never hope to understand. It may be that physics is founded on a number of brute facts, and that physicists will be driven to acknowledge it. But they are loath to assume this, and reasonably so in view of the great success of physicists in explaining the previously inexplicable. To the extent that something can be explained, it cannot be rationally opaque, and in that sense contingent.

Perhaps some things are explicable, and others not. In that case there is a limited degree of contingency, and some facts are brute facts, which can only be discovered by experiment and ob-servation. An intelligence as profound as Einstein's may be able to recognise the rational necessity for the laws of nature being Lorentz-covariant, but it may be that quantum mechanics will be forever opaque, without rhyme or reason, which we just have to accept as a brute fact about the way the world is, and shall never be able to explain or appreciate as the most fitting way for the world to be. This is a possible outcome of the quest: but it is an unpalatable one. We are reluctant to abandon the search for explanations, and settle for some things being altogether inexpli-cable which we must just learn to accept without further question. Though the verdict of experimental test is ultimately conclusive as regards fact, we should not give up our right to go on asking questions. Facts must be accepted, but should continue to be ques-tioned in our search for the explanations of things. And to allow an ultimate contingency of things in that sense would be to abandon the attempt to understand why things are as they are.

It is dangerous to speculate too closely what the future development of physics may reveal. Physicists may come up with a fundamental theory which unites all our present physical theories in one harmonious whole, whose rationale is self-evident. In that case there will be no ultimately contingent physical laws, but everything will be rationally transparent, and the only real contingencies will lie in the disposition of initial conditions or the way probabilistic laws work out in the event. Or it may be more complicated than that. Our concept of explanation is multiform and open-ended: there are many different sorts of explanation, which have been, or are yet to be, discovered or evolved. The concept of contingency is correspondingly multiform. At every stage in physics there have been contingencies so far as currently recognised types of explanation are concerned: but it has then become the task of physics to explain them, a task not always unsuccessfully attempted. Whether ultimate success will for ever elude us, or whether we shall in the end come to the end of our quest and have achieved an ultimate explanation of all physical phenomena, is a question we not only cannot answer now, but cannot even fully understand. We do not know what we are looking for, or what or what an ultimate explanation would be like. But much of physics is devoted to that search, and rests upon the faith that the answer, whatever surprises are in store for us along the way, will turn out to be explanatory and ultimately rational.

Author Index

Subject Index